Xylanolytic Enzymes

Xylanolytic Enzymes

Pratima Bajpai

AMSTERDAM • BOSTON • HEIDELBERG • LONDON
NEW YORK • OXFORD • PARIS • SAN DIEGO
SAN FRANCISCO • SINGAPORE • SYDNEY • TOKYO
Academic Press is an imprint of Elsevier

Academic Press is an imprint of Elsevier
The Boulevard, Langford Lane, Kidlington, Oxford, OX5 1GB, UK
225 Wyman Street, Waltham, MA 02451, USA

First published 2014

Notices
Knowledge and best practice in this field are constantly changing. As new research and experience broaden our understanding, changes in research methods, professional practices, or medical treatment may become necessary.

Practitioners and researchers must always rely on their own experience and knowledge in evaluating and using any information, methods, compounds, or experiments described herein. In using such information or methods they should be mindful of their own safety and the safety of others, including parties for whom they have a professional responsibility.

To the fullest extent of the law, neither the Publisher nor the authors, contributors, or editors, assume any liability for any injury and/or damage to persons or property as a matter of products liability, negligence or otherwise, or from any use or operation of any methods, products, instructions, or ideas contained in the material herein.

British Library Cataloguing in Publication Data
A catalogue record for this book is available from the British Library

Library of Congress Cataloging-in-Publication Data
A catalog record for this book is available from the Library of Congress

ISBN: 978-0-12-801020-4

For information on all Academic Press publications
visit our website at **store.elsevier.com**

This book has been manufactured using Print On Demand technology. Each copy is produced to order and is limited to black ink. The online version of this book will show color figures where appropriate.

Working together
to grow libraries in
developing countries

www.elsevier.com • www.bookaid.org

CONTENTS

Xylan is the principal type of hemicellulose. It is a linear polymer of β-D-xylopyranosyl units linked by (1−4) glycosidic bonds. In nature, the polysaccharide backbone may be added to 4-O-methyl-α-D-glucuronopyranosyl units, acetyl groups, α-L-arabinofuranosyl, and others in variable proportions. An enzymatic complex is responsible for the hydrolysis of xylan, but the main enzymes involved are endo-1,4-β-xylanase, and β-xylosidase. These enzymes are produced by fungi, bacteria, yeast, marine algae, protozoans, snails, crustaceans, insect, seeds, etc., but the principal commercial source is filamentous fungi. Recently, there has been much industrial interest in xylan and its hydrolytic enzymatic complex: as a supplement in animal feed; for the manufacture of bread, food, and drinks, textiles, pulp, and paper; ethanol and xylitol production; and as a fermentation substrate for the production of enzymes. This book describes: xylan as a major component of plant hemicelluloses; xylan's occurrence and structure and its interaction with plant cell walls; properties, production, purification, and immobilization of enzymes and their industrial application; and multiple forms of xylanases and the synergism between the enzymes of the xylanolytic complex.

CHAPTER 1

Introduction

Xylanases (EC 3.2.1.8) catalyze the hydrolysis of xylan, which is the second-most abundant polysaccharide and a major component in plant cell walls. These compounds are present in the cell wall and in the middle lamella of plant cells. This term covers a range of noncellulose polysaccharides composed, in various proportions, of monosaccharide units such as D-xylose, D-mannose, D-glucose, L-arabinose, D-galactose, D-glucuronic acid, and D-galacturonic acid. Classes of hemicellulose are named according to the main sugar unit. Thus, when a polymer is hydrolyzed and yields xylose, it is a xylan; in the same way, hemicelluloses include mannans, glucans, arabinans, and galactans (Whistler and Richards, 1970; Viikari et al., 1994; Uffen, 1997; Ebringerova and Heinze, 2000).

In nature, wood hemicelluloses rarely consist of just one type of sugar. Usually they are complex structures made of more than one polymer: the most common being glucuronoxylans, arabinoglucuronoxylans, glucomannans, arabinogalactans, and galactoglucomannans (Haltrich et al., 1996; Sunna and Antranikian, 1997; Kulkarni et al., 1999; Subramaniyan and Prema, 2002). The amount of each component varies from species to species and even from tree to tree. Therefore, hemicellulose is not a well-defined chemical compound, but a class of polymer components of plant fibers, with properties peculiar to each one. Hemicelluloses mainly comprise xylans, which are degraded by xylanolytic enzymes.

These enzymes are produced mainly by microorganisms: bacteria, fungi, actinomycetes, and yeast (Hrmova et al., 1984; Liu et al., 1998, 1999; Gilbert and Hazlewood, 1993; Sunna and Antranikian, 1997; Ball and McCarthy, 1989; Bajpai, 1997, 2009; Beg et al., 2000). The enzymes take part in the breakdown of plant cell walls—along with other enzymes that hydrolyze polysaccharides—and also digest xylan during the germination of some seeds, for example, in the malting of barley grain. Xylanases also can be found in marine algae, protozoans, crustaceans, insects, snails, and seeds of land plants (Sunna and Antranikian, 1997).

Among microbial sources, filamentous fungi are especially interesting as they secrete these enzymes into the medium and their xylanase levels are very much higher than those found in yeasts and bacteria. Many of the microorganisms producing xylanases are saprotrophs, requiring these enzymes for plant degradation and liberation of xylose, a primary carbon source for cell metabolism. Others are plant pathogens requiring hemicellulose degradation for plant cell infection. Common well-studied xylan degrading organisms are, for instance, *Trichoderma* or *Aspergillus* species. Most of their xylan-degrading enzymes are identified, characterized, and also expressed in other xylanase negative organisms such as *Escherichia coli* or *Saccharomyces cerevisiae*. Interest in xylanases from different sources has increased markedly in the last two decades. Several patents have been issued for various xylanase sources, uses, or production methods.

Xylanases are of interest for several reasons. They are clearly involved in providing sources of carbon and energy for the organisms that produce them. They act in the same fashion for hosts harboring xylanase-producing organisms, and they are involved in the growth, maturation, and ripening of cereals and fruits. Moreover, xylanases appear to be involved in the invasion of plants and fruits by pathogens. In addition, xylan-degrading enzyme systems have great potential in several biotechnological applications (Bajpai, 1997). Diverse forms of these enzymes exist, displaying varying folds, mechanisms of action, substrate specificities, hydrolytic activities, and physicochemical characteristics. Research has mainly focused on two of the xylanase-containing glycoside hydrolase families, namely, families 10 and 11, yet enzymes with xylanase activity belonging to families 5, 7, 8, and 43 have also been identified and studied, albeit to a lesser extent (Collins et al., 2005). An increasing number of reports and articles mentioning the isolation of newer microbial species for xylanase production reveal an ever-increasing interest by the scientific community in this field.

Xylanase enzymes have a wide range of industrial applications (Woodward, 1984; Biely, 1985; McCleary, 1986; Viikari et al., 1986, 1994; Sharma, 1987; Fengler and Marquardt, 1988; Poutanen and Puls 1988; Groot Wassink et al., 1989; Linko et al., 1989; Pettersson and Aman, 1989; Biely et al., 1991; Campbell et al., 1991; Zeikus et al., 1991; Bedford and Classen, 1992, 1993; Grahm and Inborr, 1992; Maat et al., 1992; Wong and Saddler, 1992a, b; Kuhad and Singh, 1993;

Sprössler, 1997; Bajpai, 1999, 2009; Puchart et al., 1999; Kapoor et al., 2001; Keskin et al., 2004; Harris and Ramalingam, 2010). There has been much industrial interest in xylanases for wood pulp biobleaching, papermaking, the manufacture of food and beverages, animal nutrition, and bioethanol production. Because of their biotechnological characteristics, xylanases are most often produced from microorganisms for commercial applications. The most promising industrial uses of xylanases are described below.

Xylanases are used in the biobleaching of wood pulp and in the bioprocessing of textiles. In fact, treating cellulosic pulps with xylanases selectively removes residual xylan when dissolving pulp. Xylanases are able to degrade the hemicellulose present in the pulp without affecting the cellulose. Enzymatic treatment has been shown to enhance various physical properties of paper, including viscosity, tensile strength, breaking length, and tear factor. In addition, biobleaching with xylanases softens the fibers, allowing them to undergo further chemical bleaching.

Another industrial use of xylanases is the bioconversion of xylan into higher value-added products. As the enzymatic hydrolysis of xylan leads to xylose, different fermentations may occur and a variety of compounds can result from these reactions. One of the most important such products is xylitol, which is used to sweeten food products such as chewing gum, candy, soft drinks, and ice cream. Xylitol can also be used as a natural sweetener in toothpaste and various pharmaceutical products. Furthermore, xylanases are used in manufacturing animal feed. Animal science researchers have proven that the pretreatment of agricultural silage and grain feed with xylanases improves nutritional value and facilitates digestion in ruminants. It has also been reported that xylanases reduce viscosity and increase absorption by breaking down the starch polysaccharides in high-fiber rye- and barley-based feeds. Supplementing broiler diets with a combination of xylanases increases growth performance indicators like weight gain.

In the bioenergy industry, there is a type of xylanase, called "Xtreme" xylanase, which has great potential to revolutionize biorefining. Xtreme xylanase is the most thermal- and acid-stable xylanase ever discovered, meaning it can tolerate a very wide range of processing conditions. Scientists report that Xtreme xylanase can be used to improve biomass pretreatment economics by removing or reducing the need for steam and pH neutralization during the biorefining process (US Department of Energy, 2011).

Beyond these benefits, Xtreme xylanase is also able to boost fermentation efficiency and, more generally, make bioprocessing more economical. Xylanases are also used for degumming of plant fiber sources such as flax, hemp, jute, and ramie and for macerating plant cell walls. The market for xylanases has been growing since the beginning of the 1990s. Commercial xylanases are industrially produced, for example, in Japan, Finland, Germany, Republic of Ireland, Denmark, Canada, and the US. The main microorganisms used to obtain these enzymes are *Aspergillus niger*, *Trichoderma sp.*, and *Humicola*. Several commercial xylanases with different properties are now available from enzyme producers (Table 1.1). With the development and application of molecular biology, structural biology, and protein engineering, significant progress has been made in the research on structures and functions of xylanases. The research progress and trend in the structure correlating with the important properties of xylanases have been reviewed (Yang et al., 2005). Analyses of three-dimensional structures and properties of mutants has revealed that glutamine and aspartic acid residues are involved in the catalytic mechanism. The thermostability of xylanases correlates with many factors, such as disulfide

Table 1.1 Some Commercially Available Xylanases	
Name of the Company	**Name of the Product**
Biocon India	Bleachzyme F
Thomas Swan Co., UK	Ecozyme
Nalco-Genencor, Ciba, -Geigy	Irgazyme 40
Solvay Enzymes	Solvay pentonase
Stern-Enzym	Sternzym HC 46
Novozymes	Pulpzyme HC, NS 51024, NS 51025
Verenium	Luminase PB-100
Advanced Enzyme	SEBrite BL 1
Dyadic	FibreZyme LBL CONC, FibreZyme PBL 100
Genencor	Optimase CX 72 L, Multifect XL
AB Enzymes	Ecopulp
Shin Nihon	Sumizyme X
Iogen Bio-Products Corporation	Biobrite 100 series

bridges, salt bridges, aromatic interactions, and content of arginine and proline; and some multidomain xylanase have thermostability domains in N- or C-terminal. But no single mechanism is responsible for the remarkable stability of xylanases. The isoelectric points and reaction pH of xylanase are influenced by hydrophobicity and content of electric charges. Many researchers have demonstrated that aromatic amino acid, histidine, and tryptophan play an important role in improving enzyme−substrate affinity. The research on structures and functions of xylanases are of great significance in understanding the catalytic mechanism and directing the improvement of properties of xylanases to meet application requirements.

REFERENCES

Bajpai, P., 1997. Microbial xylanolytic enzyme system: properties and applications. Adv. Appl. Microbiol. 43, 141−194.

Bajpai, P., 1999. Application of enzymes in the pulp and paper industry. Biotechnol. Progr. 15, 147−157.

Bajpai, P., 2009. Xylanases. In: Schaechter, M., Lederberg, J. (Eds.), Encyclopedia of Microbiology, 3rd edition. Academic Press, San Diego, pp. 600−612.

Ball, A.S., McCarthy, A.J., 1989. Saccharification of straw by actinomycete enzyme. J. Appl. Bacteriol. 66, 439−444.

Bedford, M.R., Classen, H.L., 1992. The influence of dietary xylanase on intestinal viscosity and molecular weight distribution of carbohydrates in rye-fed broiler chick. In: Visser, J., Beldman, G., VanSomeren, M., Voragen, A. (Eds.), Xylans and Xylanases. Elsevier, Amsterdam.

Bedford, M.R., Classen, H.L., 1993. An in vitro assay for prediction of broiler intestinal viscosity and growth when fed rye-based diets in the presence of exogenous enzymes. Poult. Sci. 72, 137−143.

Beg, Q.K., Bhushan, B., Kapoor, M., Hoondal, G.S., 2000. Production and characterization of thermostable xylanase and pectinase from a Streptomyces sp. QG-11-3. J. Ind. Microbiol. Biotechnol. 24, 396−402.

Biely, P., 1985. Microbial xylanolytic systems. Trends Biotechnol. 3, 286−290.

Biely, P., Vrsanska, M., Claeyssens, M., 1991. The endo-1,4-beta-glucanase I from Trichoderma reesei. Action on beta-1,4-oligomers and polymers derived from D-glucose and D-xylose. Eur. J. Biochem. 200, 157−163.

Campbell, G.L., Teitge, D.A., Classen, H.L., 1991. Genotypic and environmental differences in rye fed to broiler chicks with dietary pentosanase supplementation. Can. J. Anim. Sci. 71, 1241.

Collins, T., Gerday, C., Feller, G., 2005. Xylanases, xylanase families and extremophilic xylanases. FEMS Microbiol. Rev. 29, 3−23.

Ebringerová, A., 2005. Structural diversity and application potential of hemicelluloses. Macromol. Symp. 232 (1), 1−12.

Ebringerova, A., Heinze, T., 2000. Xylan and xylan derivatives − Biopolymers with valuable properties, 1 − Naturally occurring xylans structures, procedures and properties. Macromolecular Rapid Communications 21 (9), 542−556.

Fengler, A.I., Marquardt, R.R., 1988. Water-soluble pentosans from rye: II. Effects on rate of dialysis and retention of nutrients by the chick. Cereal Chem. 65, 298–302.

Gilbert, H.J., Hazlewood, G.P., 1993. Bacterial cellulases and xylanases. J. Gen. Microbiol. 139, 187–194.

Grahm, H., Inborr, J., 1992. Xylans and xylanases. In: Visser, J., Beldman, G., Kusters-van Someren, M.A., Voragen, A.G.J. (Eds.), Progress in Biotechnology, 7. Elsevier Science, Amsterdam, p. 539.

Groot Wassink, J.W.D., Campbell, G.L., Classen, H.L., 1989. Fractionation of crude pentosanase (arabinoxylanase) for improvement of the nutritional value of rye for broiler chickens. J. Sci. Food Agric. 46, 289.

Haltrich, D., Nidetzky, B., Kulbe, K.D., Steiner, W., Zupancic, S., 1996. Production of fungal xylanases. Bioresour. Technol. 58, 137–161.

Harris, A.D., Ramalingam, C., 2010. Xylanases and its application in food industry: a review. J. Exp. Sci. 1 (7), 1–11.

Hrmova, M., Biely, P., Vrsanska, M., Petrakova, E., 1984. Induction of cellulose- and xylan-degrading enzyme complex in the yeast Trichosporon cutaneum. Arch. Microbiol. 138, 371–376.

Kapoor, M., Beg, Q.K., Bhushan, B., Singh, K., Dadhich, K.S., Hoondal, G.S., 2001. Application of an alkaline and thermostable polygalacturonase from Bacillus sp. MG-cp-2 in degumming of ramie *(Boehmeria nivea)* and sunn hemp *(Crotalaria juncea)* bast fibers. Proc. Biochem. 36, 803–807.

Keskin, S.O., Sumnu, G., Sahin, S., 2004. Usage of enzymes in a novel baking process. Nahrung/Food 48, 156–160.

Kuhad, R.C., Singh, A., 1993. Lignocellulosic biotechnology: current and future prospects. Crit. Rev. Biotechnol. 13, 151–172.

Kulkarni, N., Shendye, A., Rao, M., 1999. Molecular and biotechnological aspects of xylanases. FEMS Microbiol. Rev. 23, 411–456.

Linko, M., Poutanen, K., Viikari, L., 1989. In: Coughlan, M.P. (Ed.), Enzyme Systems for Lignocellulose Degradation. Elsevier. Applied Science, London, p. 331.

Liu, W., Zhu, W., Lu, Y., Kong, Y., Ma, G., 1998. Production, partial purification and characterization of xylanase from Trichosporon cutaneum SL 409. Proc. Biochem. 33, 331–336.

Liu, W., Lu, Y., Ma, G., 1999. Induction and glucose repression of endo-β-xylanase in the yeast Trichosporon cutaneum SL409. Proc. Biochem. 34, 67–72.

Maat, J., Roza, M., Verbakel, J., Stam, H., daSilra, M.J.S., Egmond, M.R., et al., 1992. Xylanases and their application in bakery. In: Visser, J., Beldman, G., vanSomeren, M.A.K., Voragen, A.G.J. (Eds.), Xylans and Xylanases. Elsevier, Amsterdam, pp. 349–360.

McCleary, B.V., 1986. Enzymatic modification of plant polysaccharides. Int. J. Biol. Macromol. 8, 349–354.

Pettersson, D., Aman, P., 1989. Enzyme supplementation of a poultry diet containing rye and wheat. British J. Nutr. 62, 139–149.

Poutanen, K., Puls, J., 1988. Characteristics of Trichoderma reesei beta-xylosidase and its use in the hydrolysis of solubilized xylans. Appl. Microbiol. Biotechnol. 28, 425.

Puchart, V., Katapodis, P., Biely, P., Kremnicky, L., Christakopoulos, P., Vrsanska, M., et al., 1999. Production of xylanases, mannanases, and pectinases by the thermophilic fungus Thermomyces lanuginosus. Enzyme Microb. Technol. 24, 355–361.

Sharma, H.S.S., 1987. Enzymatic degradation of residual non-cellulosic polysaccharides present on dew-retted flax fibers. Appl. Microbiol. Biotechnol. 26, 358–362.

Sprössler, B.G., 1997. Xylanases in baking. In: Angelino, S.A.G.F., Hamer, R.J., van Hartingsveldt, W., Heidekamp, F., van der Lugt, J.P. (Eds.), Proceedings of the 1st European Symposium on Enzymes and Grain Processing (ESEGP–1). Zeist (The Netherlands): TNO Nutrition and Food Research Institute, pp. 177–187.

Subramaniyan, S., Prema, P., 2002. Biotechnology of microbial xylanases: enzymology, molecular biology and application. Crit. Rev. Biotechnol. 22, 33–46.

Sunna, A., Antranikian, G., 1997. Xylanolytic enzymes from fungi and bacteria. Crit. Rev. Biotechnol. 17, 39–67.

Uffen, R.L., 1997. Xylan degradation: a glimpse at microbial diversity. J. Ind. Microbiol. Biotechnol. 19, 1–6.

US Department of National Energy, 2011. Xtreme Xylanase Discovery Aims to Revolutionize Biorefineries. Idaho National Library. Idaho National Library. Available from <www.inl.gov/ research/xtreme-xylanase/d/xtreme-xylanase.pdf>.

Viikari, L., Ranua, M., Kantelinen, A., Sundquist J., Linko, M., 1986. Bleaching with enzymes. Proceedings of the 3rd International Conference on Biotechnology in Pulp and Paper Industry. Stockholm, Sweden, p. 67.

Viikari, L., Kantelinen, A., Sundquist, J., Linko, M., 1994. Xylanases in bleaching: from an idea to industry. FEMS Microbial. Rev. 13, 335–350.

Whistler, R.L., Richards, E.L., 1970. Hemicelluloses. In: Pigman, W., Horton, D. (Eds.), The Carbohydrates, second ed. Academic Press. pp. 447–469.

Wong, K.K.Y., Saddler, J.N., 1992a. Trichoderma xylanases: their properties and purification. Crit. Rev. Biotechnol. 12, 413–435.

Wong, K.K.Y., Saddler, J.N., 1992b. Trichoderma xylanases: their properties and application. In: Visser, J., Beldman, G., Someren, M.A.K., Voragen, A.G.J. (Eds.), Xylans and Xylanases. Elsevier, Amsterdam, pp. 171–186.

Woodward, J., 1984. Xylanases: functions, properties and applications. Topics Enz. Ferment. Biotechnol. 8, 29–30.

Yang, H.M., Yao, B., Fan, Y.L., 2005. Recent advances in structures and relative enzyme properties of xylanase. FEMS Microb. Rev. 21 (1), 6–11.

Zeikus, J.G., Lee, C., Lee, Y.E., Saha, B.C., 1991. Thermostable saccharidases. New sources, uses and biodesign. ACS Symp. Ser. 460, 36.

CHAPTER 2

Xylan: Occurrence and Structure

The substrate of xylanases, xylan, is a major structural polysaccharide in plant cells. It is found in the cell walls of land plants, in which it may constitute more than 30% of the dry weight. Aside from terrestrial plants, in which xylans are based on a ß-1,4-linked D-xylosyl backbone, marine algae also synthesize xylans of different chemical structure based on a ß-1,3-linked D-xylosyl backbone. In some species of *Chlorophyceae* and the *Rhodophyceae* in which cellulose is absent, xylans form a highly crystalline fibrillar material. Xylan structure is variable, ranging from linear 1,4-ß-linked polyxylose sugars to highly branched heteropolysaccharides. Some major structural features are summarized in Figure 2.1. The main chain of xylan is analogous to that of cellulose, but composed of D-xylose instead of D-glucose. Branches consist of L-arabinofuranose linked to the 0−3 positions of D-xylose residues and of D-glucuronic acid or 4-O-methyl-D-glucuronic acid linked to the 0−2 position. Both side-chain sugars are linked α-glycosidically. The degree of branching varies depending on the source. Xylans of several wood species—particularly of hardwoods—are acetylated. For example, birch xylan contain >1 mol of acetic acid per 2 mol of D-xylose. Acetylation occurs more frequently at the 0−3 than the 0−2 position, and double acetylation of a D-xylose unit has also been reported. The main structural elements commonly found in land plant cell wall xylans are shown in Table 2.1.

Hemicellulosic material constitutes around 30−35% of hardwood, 15−30% of graminaceous plants, and 7−12% of gymnosperms (Whistler and Richards, 1970; Wong et al., 1988; Georis et al., 2000; Bajpai, 1997, 2009). Schulze (1981) first used the term "hemicellulose" for the fractions that he collected from plant material isolated with diluted alkali. Plant biomass, in terms of dry weight, comprises three major polymeric constituents: (i) cellulose, an insoluble polymer composed of ß-D-glucopyranosyl residues linked by ß-1,4-glycosidic bonds; (ii) hemicellulose, a series of heteropolysaccharides that includes xylans, glucans, mannans, and arabinans; and (iii) lignin, a complex polyphenol,

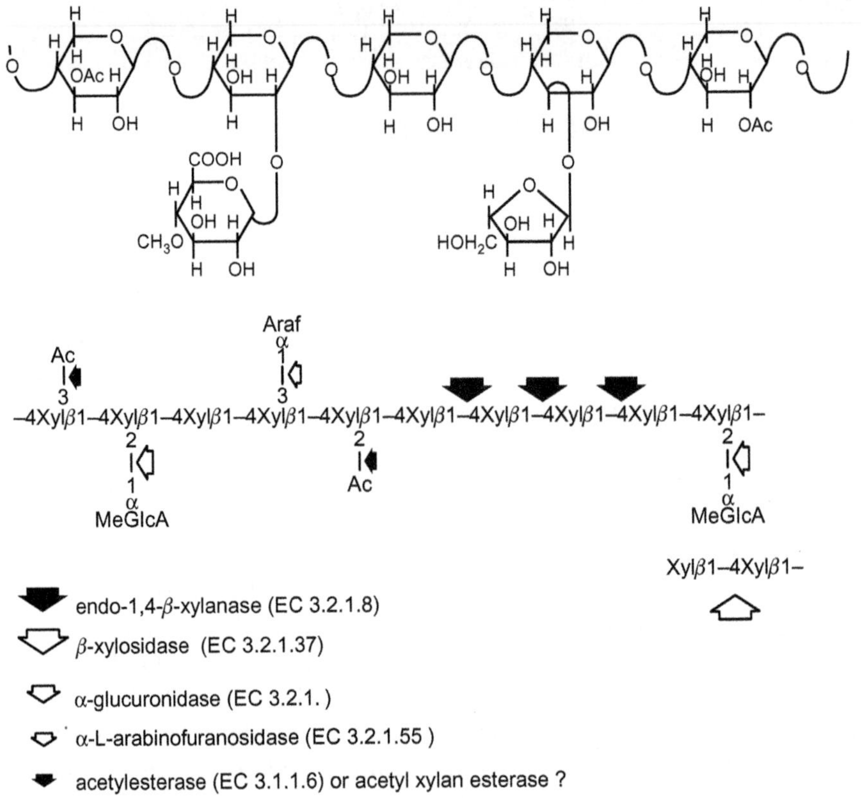

Araf
α
1

Ac
|
3

$-4Xyl\beta1-4Xyl\beta1-4Xyl\beta1-4Xyl\beta1-4Xyl\beta1-4Xyl\beta1-4Xyl\beta1-4Xyl\beta1-4Xyl\beta1-4Xyl\beta1-$

2
1
α
MeGlcA

2
1
Ac

2
1
α
MeGlcA

$Xyl\beta1-4Xyl\beta1-$

endo-1,4-β-xylanase (EC 3.2.1.8)

β-xylosidase (EC 3.2.1.37)

α-glucuronidase (EC 3.2.1.)

α-L-arabinofuranosidase (EC 3.2.1.55)

acetylesterase (EC 3.1.1.6) or acetyl xylan esterase ?

Figure 2.1 A hypothetical plant xylan and the sites of its attack by microbial xylanolytic enzymes (Reproduced from Biely (1985), copyright (1985), with permission from Elsevier Science.)

intimately interconnected with the hemicelluloses, forming a matrix that surrounds the orderly cellulose microfibrils. In wood, lignin in high concentration is the glue that binds contiguous cells. As a whole, biomass comprises on average 23% lignin, 40% cellulose, and 33% hemicellulose by dry weight. It does not accumulate in nature, but undergo microbial degradation as part of the carbon cycle (Biely, 1993).

The main heteropolymers of the hemicellulosic component are xylan, mannan, galactans, and arabinans. D-xylose, D-mannose, D-galactose, and L-arabinose are examples of sugar moieties that are commonly attached with the heteropolymers and on the basis of which these heteropolymers are classified. Xylan molecules are mainly constituted by D-xylose as the monomeric unit, and traces of L-arabinose are also present (Bastawde, 1992). Further, some substituents, such as acetyl, arabinosyl, and glucuronysyl, are found on the backbone of xylan (Whistler

Table 2.1 Main Structural Elements Commonly found in Land Plant Cell Wall Xylans

Structural Type	Source
Linear homoxylan	Esparto grass Tobacco stalk

Arabinoxylan
  Low branching
  High branching
  Complex side chain
	
	Common barberry monocots
	Primary walls
	Flours
	Gramineae pericarp

Glucuronoxylan	Soybean hull Hardwood Gramineae Legumes
Glucuronoarabinoxylan	Softwoods Gramineae Dicot primary walls

Based on Chanda et al., 1950; Timell, 1965; Eda et al., 1976; Wilkie, 1979; Bajpai, 1997

and Richard, 1970). Xylan occupies the central position inbetween the sheath of lignin residues and also covalently linked and intertwined with this sheath at several points. Covalent linkage of xylan with lignin sheath and interwinedness through inter H-bonding gives an appearance of a "coat" around the cellulosic monomers (Biely, 1985; Joseleau et al., 1992). Intra-chain H-bonding occurs through the $0-3$ position, giving unsubstituted xylan a helical twist.

These cellulose monomers will act as a barrier against the hydrolyzing action of cellulose enzyme (Uffen, 1997).

The xylan in hardwood, which may account for $10-35\%$ of the total dry weight, is acetyl-4-O-methylglucuronoxylan with a degree of polymerization (DP) between 150 and 200. Approximately one in ten of the ß-D-xylopyranose backbone units is substituted at C-2 with a 1,2-linked 4-O-methyl-ß-Dglucuronic acid residue, while 70% is acetylated at C-2, C-3, or both. In fact, birch xylan contains more than 1 mol of acetic acid per 2 mols of xylose. The presence of these acetyl groups is responsible for the partial solubility of xylan in water. These groups are readily removed when xylan is subjected to alkali extraction. Softwoods contain $10-15\%$ xylan as arabino-4-O-methylglucuronoxylan with a DP less than that of hardwood xylans, between 70 and 130. This material is not acetylated and contains ß-D-xylopyranose, 4-O-methyl-β-Dglucuronic acid and L-arabinofuranose in a ratio of 100:20:13. The 4-O-methylglucuronose residues are attached to C-2 and the L-arabinofuranose residues linked by α-1,3 glycosidic bonds to the C-3 of the relevant xylopyranose backbone units. The arabinoyl constituents occur on almost 12% of the xylosyl residues. The ester-linked components (acetyl, feruloyl, and p-coumaroyl residues) may be lost from substrates prepared by solubilization in alkali. In native lignocellulolytic material, some or all of the feruloyl substituents may be involved in covalent cross-linking with other polysaccharides. However, additional natural functions of these phenolic side groups have been suggested, for example, regulation of cell-wall extension, stabilization, and defense against invading plant pathogens (Tenkanen and Poutanen, 1992). Homoxylans consisting exclusively of xylosyl residues are not widespread in nature; they have been isolated from a few sources like esparto grass, tobacco stalks, and guar seed husk (Sunna and Antranikian, 1997). However, on the basis of the nature of the substituents, a broad distinction may therefore be made among xylans in which complexity increases from linear to highly substituted xylans. There are four main families of xylan: Arabinoxylans, Glucuronoxylans, Glucuronoarabinoxylans, and Galactoglucuronoarabinoxylans (Voragen et al., 1992).

Arabinoxylans have been identified in wheat, rye, barley, oat, rice, and sorghum, as well as in some other plants: pangola grass, bamboo shoots, and rye grass. Although these polysaccharides are minor components of entire cereal grains, they constitute an important part of plant cell walls (Izydorczyk and Biliaderis, 1995). Glucuronoxylans and glucuronoarabinoxylans are located mainly in the secondary wall and function as an adhesive by forming covalent and noncovalent bonds with lignin, cellulose, and other polymers essential to the integrity of the cell wall. Xylans are the principal class of hemicelluloses in angiosperms contributing 15–30% of the total dry weight, but are less abundant in gymnosperms, which contain 7–12% xylans (Haltrich et al., 1996). Glucuronoxylans are linear polymers of β-D-xylopyranosyl units linked by (1–4) glycosidic bonds (xylose). This polysaccharide backbone has 4-O-methyl-α-D-glucuronopyranosyl units, D-glucuronosyl units methylated at position 4 and joined to position 2 or 3 of the β-D-xylopyranosyl. Angiosperm (hardwood) glucuronoxylans also have a high rate of substitution (70–80%) by acetyl groups, at position 2 and/or 3 of the β-D-xylopyranosyl, conferring on the xylan its partial solubility in water (Coughlan and Hazlewood, 1993).

On the basis of the nature of the substituents, a broad distinction may be made among xylans in which complexity increases from linear to highly substituted xylans. Four main families can be considered (Voragen et al., 1992): Arabinoxylans having only side chains of single terminal units of α-L-arabinofuranosyl substituents, and—in the particular case of cereals—arabinoxylans varying in the degree of arabinosyl substitution either 2-O- and 3-O-mono-substituted or double (2-O-, 3-O-) substituted xylosyl residues. Glucuronoxylans in which α-D-glucuronic acid or its 4-O-methyl ether derivative represents the only substituent. In Glucuronoarabinoxylan, α-D-glucuronic (and 4-O-methyl-α-Dglucuronic) acid and α-L-arabinose are present at the same time. Galactoglucuronoarabinoxylans are characterized by the presence of terminal ß-D-galactopyranosyl residues on complex oligosaccharide side chains of xylans. These are typically found in perennial plants. In this area, specific degradation involving chemical reagents can provide useful results on structural analysis, but certainly the best tools are the enzymes. There is a great diversity of available xylan-degrading enzymes, glycohydrolases, and glycanohydrolases, which, when used in conjunction with modern techniques for the analysis of oligosaccharides like HPLC, NMR, or mass spectroscopy, yield the best results for the detailed description of the true primary structure of xylans. In a pioneer work, Vliegenthat et al. (1992) have carried out a study showing that a breakthrough could be achieved in the structure elucidation of wheat arabinoxylans. The approach consists of: degradation of polymer by endo-xylanase to a family of oligosaccharides, separation and purification of this mixture to single compounds, and, finally, identification of the oligosaccharides by NMR spectroscopy. It could be shown that this approach is also suitable for defining the substrate specificity of endoxylanases in terms of their structural requirements.

For complete extraction of xylans from secondary cell walls, drastic alkaline conditions are needed, suggesting strong bonds between xylans and other wall polymers. There are numerous pieces of evidence suggesting that different types of covalent bonds maintain an association between lignin and xylan in the secondary cell wall, most of them being susceptible to alkaline cleavage. These covalent bonds involve ester linkages between glucuronoxylans and lignin via benzyl ester bonds with the carboxylate group of 4-O-methylglucuronic acid. The second-most reported type of covalent bonds between xylan and lignin are ether linkages that involve L-arabinofuranose side chains or xylose units. Other covalent linkages likely to interconnect xylan with lignin in the cell wall are oxygen-containing bonds, acetal, and glycosides (Joseleau et al., 1992). In similar types of covalent interactions, xylan chains may be interconnected with other xylan chains or other polymers. It has been demonstrated that ferulic acid and p-coumaric acid bridges exist between arabinoxylan segments and it has been suggested that galacturonans could be covalently linked to arabinoxylans. Several possibilities for covalent interactions between xylans and other wall polysaccharides, particularly pectin, have also been suggested (Selvendran, 1985). Although there is no direct evidence for covalent linkages between cellulose and other components of the cell wall, xylans are believed to hydrogen bond to cellulose microfibrils with a strength that could be inversely proportional to the degree of side-chain substitution. Joseleau et al. (1992) suggest that the capacity of xylans to associate with cellulose depends upon their conformation in situ in the cell wall relative to that of cellulose.

High rates of acetylation at C-2 and C-3 atoms of hardwood xylan make it partially soluble in water. These rates also are responsible for the easy exit of acetyl groups through alkali treatment (Sunna and Antranikian, 1997). Unlike hardwood xylans, softwood xylans are not acetylated and are freely soluble in water. Further, they have α-L-arabinofuranose units linked through α-1,3-glycosidic bonds at the C-3 position (Puls and Schuseil, 1993), are shorter in length compared to the hardwood xylan, and also possess a lower degree of branching. Contrary to this, homoxylans are restricted to esparto grass (Chanda et al., 1950) and tobacco stalks (Eda et al., 1976) and only consist of xylosyl residues. Degradability of native xylan is minimal; it is present as a formidable substrate, whereas alkaline extraction of the native form de-esterifies the substrate and is responsible for the removal of acetyl groups and breakdown of cross linkages, hence increasing the enzymatic degradability of xylan. Due to all these properties of commercial xylan, it is preferred over the native form for assay methods and screening procedures. Extraction of highly purified xylan is not achievable, as it contains some inseparable substituents. Therefore, several aspects of xylan structure remain unclear.

Xylans are the main hemicelluloses in hardwood and they also predominate in annual plants and cereals, making up to 30% of the cell wall material. They are also one of the major constituents (25−35%)

of lignocellulosic materials. The potential sources of xylans include many agricultural crops such as straw, sorghum, sugar cane, corn stalks, and cobs, and hulls and husks from starch production, as well as forest and pulping waste products from hardwoods and softwoods (Ebringerova and Heinze, 2000; Kayserilioglu et al., 2003).

There is a relationship between the chemical structure of xylans and their botanical origin and also in their cytological localization. This results in a certain degree of complexity of xylan-containing materials that may possess several different xylan polymers of related structures but differ by more or less important features. Xylans of terrestrial plants are present in various proportions in the cell wall of all lignified tissues, but many may also be found in plant species as diverse as mosses and ferns. They usually are constituents of the secondary walls of tissues having structural functions, but they are also present to some extent in the primary walls of growing cells as well as in the primary walls of seeds and bulbs of certain plant species in which they have reserve functions. In all of the various ultrastructural localizations that they have in plant walls, xylans interact with other structural components, in particular with cellulose microfibrils, with other noncellulosic polymers, and—in most cases—with lignin. Noncovalent interactions of xylans with other polysaccharides essentially involve hydrogen bonding, whereas covalent bonds interconnect xylans, lignins, and some phenolic acids. Although relatively little is known about the conformation of xylan chains within the cell wall lattice, conformational analysis of xylans has been carried out by crystallographic studies after purification of the isolated hemicelluloses. The development of techniques that permit the observation and analysis of polymers in their natural location in the cell wall—like electron microscopy, solid-state NMR, and Fourier transform infrared spectroscopy—are now being employed.

Xylans show a large polydiversity and polymolecularity, like most polysaccharides of plant origin (Joseleau et al., 1992). This corresponds to their presence in a variety of plant species and to their distribution in several types of tissues and cells. All land plant xylans are characterized by a β-1,4-linked D-xylopyranosyl main chain that carries a variable number of neutral or uronic monosaccharide substituents or short oligosaccharide side chains (Bajpai, 1997, 2009). Very few unbranched linear xylan homopolysaccharides from land plants have been isolated. The best known is the xylan from esparto grass that, because of the peculiarity of its structure, served as a model for chemical and physical studies (Chanda et al., 1950; Marchessauk et al., 1961). Other unsubstituted linear xylans have been isolated from tobacco stalks and guar seed husks (Montgomery et al., 1956; Eda et al., 1976). From the simple β-1,4-D-xylopyranosyl chain, the structural complexity of xylan rises with the number of substituting mono- and oligosaccharides attached to the β-1,4-linked xylosyl main chain. Timell reviewed the chemistry of hemicelluloses from angiosperms and gymnosperms (Timell, 1964, 1965) and Wilkie described the structural characteristics of the different xylans from monocotyledonous species (Wilkie, 1979).

The chemical structure of alkali-soluble xylans has been known for a rather long time. Most of the chemical structures reported in the literature were obtained by chemical, enzymatic, or spectroscopic analytical methods used separately or in combination. This led to description of the most characteristic features of polysaccharides belonging to the xylan family and corresponded to averaged structures with no or little information about minor structural elements that were considered insignificant or due to impurities. The structural features of complex heteroxylans were obtained with more powerful investigative tools. These tools principally involve the use of the numerous enzymes: glycohydrolases, esterases, and glycanases and HPLC and 2D NMR spectroscopy. However, a lot less is known about their true primary structure, that is, if the arabinosyl, uronic acid, or acetyl substituents are attached to the xylosyl backbone randomly or as regular repeating sequences. On the basis of the nature of the substituents, four main families can be considered in which the complexity increases from linear to highly substituted xylans. A broad distinction may thus be made between the arabinoxylans having only side chains of single terminal units of α-L-arabinofuranosyl substituents, the true glucuronoxylans in which α-D-glucuronic acid and/or its 4-0-methyl ether derivative represent the only substituent, and the more complex glucuronoarabinoxylan in which α-L-arabinose and α-D-glucuronic acid and 4-0-methyl-α-D-glucuronic acid are present at the same time. In addition to these three main families, one may distinguish arabinoxylans having a high degree of substitution by more or less short side chains of 2,3,5- and 2,3-linked arabinofuranosyl oligosaccharides attached to the 0−3 position of the xylosyl main chain, and galactoglucurono-arabinoxylans, characterized by the presence of terminal β-D-galactopyranosyl residues on complex oligosaccharide side chains of xylans from several perennial plants (Wilkie, 1979).

REFERENCES

Bajpai, P., 1997. Microbial xylanolytic enzyme system: properties and applications. Adv. Appl. Microbiol. 43, 141–194.

Bajpai, P., 2009. Xylanases. In: Schaechter, M., Joshua Lederberg, J. (Eds.), Encyclopedia of Microbiology, Third edition. Academic Press, San Diego (California), pp. 600–612.

Bastawde, K.B., 1992. Xylan structure microbial xylanases and their mode of action. World J. Microbiol. Biotechnol. 8, 353–368.

Biely, P., 1985. Microbial xylanolytic systems. Trends. Biotechnol. 3 (11), 286–290.

Biely, P., 1993. Biochemical aspects of the production of microbial hemicellulases. In: Coughlan, M.P., Hazlewood, G.P. (Eds.), Hemicellulose and Hemicellulases. Portland Press, London, pp. 29–52.

Chanda, S.K., Hirst, E.L., Jones, J.K.N., Percival, E.G.V., 1950. The constitution of xylan from esparto grass (Stipa tenacissima). J. Chem. Soc. 50, 1287–1289.

Coughlan, M.P., Hazlewood, G.P., 1993. β-1,4-D-Xylan-degrading enzyme systems, biochemistry, molecular biology and applications. Biotechnol. Appl. Biochem. 17, 259–289.

Ebringerova, A., Heinze, T., 2000. Xylan and xylan derivatives: biopolymers with valuable properties, 1. Naturally occurring xylans structures, isolation procedures, and properties. Macromol. Rapid Commun. 21 (9), 542–556, ISSN 1022–1336.

Eda, S., Ohnishi, A., Kato, K., 1976. Xylan isolated from the stalks of Nicotiana tabacum. Agric. Biol. Chem. 40, 359–364.

Georis, J., Giannotta, F., de Buyl, E., Granier, B., Frère, J.M., 2000. Purification and properties of three endo-b-1,4-xylanases produced by Streptomyces sp. strain S38 which differ in their ability to enhance the bleaching of kraft pulp. Enzyme. Microb. Technol. 26, 178–186.

Haltrich, D., Nidetzky, B., Kulbe, K.D., Steiner, W., Zupancic, S., 1996. Production of fungal xylanases. Bioresour. Technol. 58, 137–161.

Izydorczyk, M.S., Biliaderis, C.G., 1995. Cereal arabinoxylans: advances in structure and physiochemical properties. Carbohydr. Polym. 28, 33–48.

Joseleau, J.P., Comtat, J., Ruel, K., 1992. Chemical structure of xylans and their interactions in the plant cell walls. In: Visser, J., Beldman, G., vanSomeren, M.A.K., Voragen, A.G.J. (Eds.), Xylans and Xylanases. Elsevier, Amsterdam, pp. 1–15.

Kayserilioglu, B.S., Bakir, U., Yilmaz, L., Akkas, N., 2003. Use of xylan, an agricultural byproduct, in wheat gluten based biodegradable films: mechanical, solubility and water vapor transfer rate properties. Bioresour. Technol. 87 (3), 239–246.

Marchessauk, R.H., Morehead, F.F., Walter, N.M., Glaudemans, C.P.J., 1961. Morphology of xylan single crystals. J. Polym. Sci. 51, 66–68.

Montgomery, R., Smith, F., Srivastava, H.C., 1956. Structure of cornhull hemicellulose. I. Partial hydrolysis and identification of 2-O-(α-D-glucopyranosyluronic acid)-D-xylopyranose. J. Am. Chem. Soc. 8, 2837–2839.

Puls, J., Schuseil, J., 1993. Chemistry of hemicelluloses: relationship between hemicellulose structure and enzyme required for hydrolysis. In: Coughlan, M.P., Hazlewood, G.P. (Eds.), Hemicellulose and Hemicellulases. Portland Press, London, pp. 1–28.

Schulze, E., 1981. Information regarding chemical composition of plant cell membrane. Ber. Dtsch. Chem. Ges. 24, 2277–2287.

Selvendran, R.R., 1985. Developments in the chemistry and biochemistry of pectic and hemicellulosic polymers. J. Cell. Sci. S2, 51–88.

Sunna, A., Antranikian, G., 1997. Xylanolytic enzymes from fungi and bacteria. Crit. Rev. Biotechnol. 17, 39–67.

Tenkanen, M., Poutanen, K., 1992. Significance of esterases in the degradation of xylans. In: Visser, J., Beldman, G., Kuster-van Someren, M.A., Voragen, A.G.J. (Eds.), Xylan and Xylanases. Elsevier, Amsterdam, pp. 203–212.

Timell, T.E., 1964. Wood hemicelluloses: Part I. Adv. Carbohydr. Chem. 19, 247–302.

Timell, T.E., 1965. Wood hemicelluloses. Part II. Adv. Carbohydr. Chem. 20, 409–483.

Uffen, R.L., 1997. Xylan degradation: a glimpse at microbial diversity. J. Ind. Microbiol. Biotechnol. 19, 1–6.

Vliegenthat, J.F.G., Hoffmann, R.A., Kamerling, J.P., 1992. A 1H-NMR spectroscopy study on oligosaccharides obtained from wheat arabinoxylan. In: Visser, J., Beldman, G., Kuster-van Someren, M.A., Voragen A.G.J. (Eds.), Xylan and Xylanases. pp. 17–37.

Voragen, A.G.J., Gruppen, H., Verbruggen, M.A., Vietor, R.J., 1992. Characterization of cereals arabinoxylans. In: Visser, J., Beldman, G., Kuster-van Someren, M.A., Voragen, A.G.J. (Eds.), Xylan and Xylanases. Elsevier, Amsterdam, pp. 51–67.

Whistler, R.L., Richards, E.L., 1970. Hemicelluloses. In: Pigman, W., Horton, D. (Eds.), The Carbohydrates. Academic Press, New York, pp. 447–469.

Wilkie, K.C.B., 1979. The hemicelluloses of grasses and cereals. Adv. Carb. Chem. 36, 215.

Wong, K.K., Tan, L.U., Saddler, J.N., 1988. Multiplicity of β-1,4-xylanase in microorganisms: functions and applications. Microbiol. Mol. Biol. Rev. 52, 305–317.

CHAPTER 3

Microbial Xylanolytic Systems and Their Properties

The main component of xylan is D-xylose, which is a five-carbon sugar. It can be converted to a single cell protein and chemical fuels by microbial cells (Biely, 1985). Due to the heterogeneity and complex chemical nature of plant xylan, its complete breakdown requires the action of a complex of several hydrolytic enzymes. These enzymes have diverse specificities and modes of action. The xylanolytic enzyme system carrying out the xylan hydrolysis is usually composed of several hydrolytic enzymes (Figure 2.1):

- β-1,4-endoxylanase
- β-xylosidase
- acetyl xylan esterase
- arabinase
- α-glucuronidases
- ferulic acid esterase and p-coumaric acid esterase

All of these enzymes act cooperatively to convert xylan into its constituent sugars. The presence of such a multifunctional xylanolytic enzyme system is quite widespread among fungi, actinomycetes, and bacteria (Biely et al., 1985; Dey et al., 1992; Belancic et al., 1995; Elegir et al., 1995; Bajpai, 1997, 2009, 2013). Tables 3.1–3.4 summarizes the properties of xylanases reported in literature.

3.1 XYLANASES

Endo-1,4-β-xylanases (1,4-β-D-xylan xylohydrolase; EC 3.2.1.8) are the most important xylan-degrading enzymes. These enzymes cleave the glycosidic bonds in the xylan backbone and bring a reduction in the degree of polymerization of the substrate. Xylan is not attacked randomly, but the bonds selected for hydrolysis depend on the nature of the substrate molecule (Reilly, 1981; Puls and Poutanen, 1989; Li et al., 2000), that is, on the chain length, the degree of branching, and the presence of substituents. The main hydrolysis products are

Table 3.1 Characteristics of Xylanases from Different Bacteria

Bacteria	pI	Optimum pH	Molecular Weight (kDa)
Acidobacterium capsulatum	7.3	5	41
Bacillus sp. W-1	8.5	6	21.5
Bacillus circulans WL-12	9.1	5.5–7	15
Bacillus stearothermophilus T-6	7.9	6.5	43
Bacillus sp. strain BP-23	9.3	5.5	32
Bacillus sp. strain BP-7	7–9	6.0	22–120
Bacillus polymyxa CECT 153	4.7	6.5	61
Cellulomonas fimi	4.5–8.5	5–6.5	14–150
Cellulomonas NCIM 2353	8	6.5	22,33,53
Micrococcussp. AR-135	—	7.5–9	56
Staphylococcus sp. SG-13	—	7.5–9.2	60
Thermoanaerobacterium JW/SL–YS485	4.37	6.2	24–180
Thermotoga maritima MSB8	5.6	5.4–6.2	40, 120

Based on Sunna and Antranikian, 1997; Beg et al., 2001; Subramaniyan and Prema, 2002

Table 3.2 Characteristics of Xylanases from Different Fungi

Fungi	pI	Optimum pH	Molecular Weight (kDa)
Acrophialophora nainiana	—	6	17
Aspergillus niger	9	5.5	13.5–14.0
Aspergillus kawachii	3.5–6.7	2–5.5	26–35
Aspergillus nidulans	—	5.4	22–34
Aspergillus fischeri	—	6	31
Aspergillus sojae	3.5, 3.75	5,5.5	32.7, 35.5
Aspergillus sydowii	—	5.5	30
Cephalosporium sp.	—	8	30, 70
Fusarium oxysporum	—	6	20.8, 23.5
Geotrichum candidum	3.4	4	60–70
Paecilomyces variotii	5.3	4	20
Penicillium purpurogenum	8.6, 5.9	7, 3.5	33, 23
Thermomyces lanuginosus DSM 5826	4.1	7	25.5
Thermomyces SSBP	3.8	6.5	23.6
Trichoderma harzianum	—	5	20
Trichoderma reesei	9, 5.5	5–5.5	20, 19

Based on Sunna and Antranikian, 1997; Beg et al., 2001; Subramaniyan and Prema, 2002

Table 3.3 Characteristics of Xylanases from Different Yeast			
Yeast	pH	pI	Molecular Weight (kDa)
Aureobasidium pullulans Y-2311-1	4.4	9.4	25
Cryptococcus albidus	5	—	48
Trichosporon cutaneum SL409	50	—	6.5
Based on Sunna and Antranikian, 1997; Beg et al., 2001; Subramaniyan and Prema, 2002			

Table 3.4 Characteristics of Xylanases from Different Actinomycetes			
Actinomycete	pH	pI	Molecular Weight (kDa)
Streptomyces sp. EC 10	7–8	6.8	32
Streptomyces sp.	6.7	4.8-8.3	23.8–40.5
Streptomyces OPC–520	7	4.2-8	33,547
Streptomyces T7	4.5–5.5	7.8	20
Streptomyces CECT 3336	5–8	9	48 6 50
Streptomyces viridisporus T7A	7–8	10.2–10.5	59
Thermomonospora curvata	6.8–7–8	4.2–8.4	15–36
Based on Sunna and Antranikian, 1997; Beg et al., 2001; Subramaniyan and Prema, 2002			

β-D-xylopyranosyl oligomers initially, but at a later stage, small molecules such as mono-, di-, and trisaccharides of β-D-xylopyranosyl may be produced. The endoxylanases have been classified in many ways. These enzymes can be divided into two types, according to the end-products of the reaction:

— Non-debranching enzymes: These do not hydrolyze at the 1,3-α-L-arabinofuranosyl branch-points of arabinoxylans and thus do not liberate arabinose.
— Debranching enzymes: These enzymes do hydrolyze these side-branches, liberating arabinose (Wong et al., 1988).

Both of these types have been found separately in a number of fungal species. However, there are some fungi that are capable of producing both types of xylanases, resulting in a more efficient hydrolysis of xylan. Wong et al. (1988) reported a correspondence between classes of microbial endoxylanases and their physicochemical properties, such as molecular weight (MW) and isoelectric point (pI). They were divided into two groups:

- Basic enzymes, with MW <30 kDa
- Acid endoxylanases, with MW >30 kDa

It should be pointed out, though, that this relation only works for about 70% of cases, the exact opposite being true of many known endoxylanases. Other classifications into various families, proposed for the glycosidases, was provided by Henrissat and Bairoch (1993) and by Törrönen and Rouvinen (1997). In general, the endoxylanases show peak activity between 40°C and 80°C and between pH 4.0 and 6.5, but optimal conditions have been found outside these ranges. Individual fungi and bacteria exhibit a multiplicity of endoxylanases; in some cases three or more enzyme activities have been separated from a single culture (Rizzatti et al., 2004). Several factors may be responsible for the multiple forms often detected for endoxylanases. These include differential processing of mRNA, post-translational modifications such as glycosylation and self-aggregation, and proteolytic digestion. Multiple endoxylanases can also be expressed by distinct alleles of one gene or even by completely separate genes (Sung et al., 1995; Segura et al., 1998; Chavez et al., 2002).

Nonspecific xylanases from *Trichoderma sp.* can attack cellulose, carboxymethylcellulose, p-nitrophenyl-β-glucoside, cello-oligomers, cellobiose, laminarin, and p-nitrophenyl-β-cellobioside. Carboxymethyl cellulose, p-nitrophenyl-glucoside, and xylan apparently compete for the same active site on an enzyme from *T. viride* that does not attack insoluble cellulose. A number of nonspecific glycanases have been characterized in *Trichoderma sp.* With few exceptions, they have relatively larger MWs (32−76 kDa) and more acidic pI values (3.5−5.3). This pI range excludes all of the apparently specific xylanases except for the 21-kDa xylanase from *T. lignorum*. These observations corroborate the hypothesis that there is a cluster of nonspecific glycanases with acidic pIs. The observation that these glycanases were induced by sophorose but not xylobiose suggests that they could be considered cellulases rather than xylanases. A multiplicity of xylanases has been documented in numerous organisms (Wong et al., 1988), with evidence for the occurrence of three to five xylanases in bacteria and fungi. Analyses at the molecular genetic level have verified the occurrence of multiple xylanases in bacterial species. Several xylanases have been detected in *Trichoderma sp.* and have been purified and characterized in *T. harzianum* E58. The functional and genetic basis of these multiple enzymes has not been completely elucidated. Electrophoretically distinct xylanases may arise from posttranslational modifications of the same gene product, such as differential glycosylation or proteolysis. *Trichoderma* xylanases have been reported to be glycosylated in some cases, but not in others. The latter group of enzymes includes one pair of xylanases isolated from *T. koningii* and another from *T. lignorum*. It therefore appears that differential glycosylation cannot explain the occurrence of multiple xylanases in these cases. Furthermore, a comparison of amino acid compositions suggests that the three xylanases purified from *T. harzianum* are distinct gene products (Wong et al., 1986a).

The optimal conditions for activity of *Trichoderma* xylanases range from 45°C to 65°C and from pH 3.5 to 6.5. As might be expected, the xylanases with higher temperature optima are relatively more thermally stable than those with lower temperature optima. Two xylanases have been reported to be stable at 50°C for 1 h and one at 60°C for 20 min. These properties are relatively moderate when compared to xylanases isolated from thermophilic microorganisms. For example, a xylanase isolated from *Thermostoga sp.* strain FjSS3-B.1 has a temperature optimum of 105°C at pH 5.5 and a half-life of 90 min at 95°C. Furthermore, alkaline-tolerant xylanases have been isolated from *Bacillus sp.* that have a broad range of pH optima and stabilities, ranging up to pH 10. Mercury ions have been found to be inhibitory to the activity of *Trichoderma* xylanase at concentrations ranging from 0.1 to 10 mM. One exception is a partially purified xylanase from *T. viride* that was not inhibited by 1 mM Hg. Of the other ions tested, 1 mM Ca^{2+} was found to be inhibitory in one case and 1 mM Cu^{2+} in another. *Trichoderma* xylanases have been found to be active on xylans from different sources, usually producing xylooligomers: xylobiose and xylose. Xylose is not usually the major product, and it is typically produced after an accumulation of xylooligomers. Of the xylanases characterized, one isolated from *T. pseudokoningii* and two isolated from *T. viride* were reported to be unable to produce xylose. Two nonspecific glycanases from *T. viride* were also found to be unable to produce xylose during xylan hydrolysis. One of these glycanases produced xylobiose as an initial product, indicating that it acts like an exoxylanase. The hydrolysis patterns of *Trichoderma* xylanases, however, have suggested that most are endoxylanases.

Xylans are not completely hydrolyzed by crude culture filtrates or purified xylanases from *Trichoderma sp.* However, hydrolysis yields from certain xylans could be improved by using mixtures of three different xylanases purified from *T. harzianum* (Wong et al., 1986b). Xylose yields obtained using a purified xylanase from *T. reesei* were increased when a purified β-xylosidase was added. They were further increased where the relevant debranching enzymes were added to the hydrolysis reaction. When acetylated xylooligomers were partially deactivated by freeze-drying over ammonia, they became more accessible to hydrolysis by xylanase. All of these observations suggest that the substituents on xylans can restrict their hydrolysis by xylanases.

3.2 β-D-XYLOSIDASES

ß-xylosidase is part of most microbial xylanolytic systems, but the highest extracellular production levels have been reported for fungi (Bajpai, 1997; Sunna and Antranikian, 1997; Beg et al., 2001). It has been reported that ß-xylosidases (1,4-ß-D-xyloside xylohydrolase; EC 3.2.1.37) are produced by a variety of bacteria and fungi and may be found in the culture fluid, associated with the cell, or both. These enzymes may be monomeric, dimeric, or tetrameric, with MW values ranging from 26 to 360 kDa, and may hydrolyze xylooligosaccharides and xylobiose to xylose by removing successive D-xylose residues from the nonreducing termini. ß-xylosidases are rather large enzymes with molecular weights exceeding 100 kDa and are often reported to consist of two or more subunits. Most purified ß-xylosidases show highest activity toward xylobiose and no activity toward xylan. The activity toward xylooligosaccharides generally decreases rapidly with increasing chain length. In addition to formation of xylose, many β-xylosidases produce transfer products with higher molecular weights than that of the substrate. Some ß-xylosidases have also been reported to possess ß-glucosidase activity. An important characteristic of ß-xylosidases is their susceptibility to inhibition by xylose, which may significantly affect the yield under process conditions.

Many xylosidases exhibit relaxed specificity for both the sugar and the linkage. True ß-xylosidases cleave artificial ß-xylosides and unsubstituted ß-1,4-linked xylooligosaccharides, including xylobiose. Action against xylo-oligomers proceeds with the preferential removal of xylose residues from the nonreducing end of such substrates, and affinity usually increases with decreasing DP, exhibiting little or no activity against xylan. However, in some cases the enzyme exhibits activity against xylan that is about 30% of that observed against xylobiose. ß-xylosidases may be distinguished from the less commonly found exo-xylohydrolases by virtue of the fact that the latter preferentially remove xyloses from the nonreducing end of xylans and xylo-oligomers. Thus, affinity in exo-xylohydrolase increases with increasing DP.

β-D-Xylosidases can be classified according to their relative affinities for xylobiose and larger xylooligosaccharides. Xylobiases and exo-1,4-β-xylanases can be recognized as distinct entities, as proposed by Biely (1985, 1993), but will be treated as xylosidases that hydrolyze small xylooligosaccharides and xylobiose, releasing β-D-xylopyranosyl residues from the nonreducing terminus. Purified β-xylosidases usually do not hydrolyze xylan; their best substrate is xylobiose and their affinity for xylooligosaccharides is inversely proportional to the degree of polymerization. They are able to cleave artificial substrates such as p-nitrophenyl-and o-nitrophenyl-β-D-xylopyranoside. Transxylosilation activity has also been detected in fungi, resulting in products of higher MW than the original substrates (Kurakabe et al., 1997). An important role attributed to β-xylosidases comes into play after the xylan has suffered a number of successive hydrolyses by xylanase. This reaction leads to the accumulation of short oligomers of β-D-xylopyranosyl, which may inhibit the endoxylanase. β-xylosidase then hydrolyzes these products, removing the cause of inhibition, and increasing the efficiency of xylan hydrolysis (Andrade et al., 2004; Zanoelo et al., 2004). Regarding the localization of β-xylosidases, those from filamentous fungi may be retained within the mycelium, may only be detected in cell extracts, or may be liberated into the growth medium (extracellular). For instance, the β-xylosidases from *Humicola grisea* var. *thermoidea* (Almeida et al., 1995) and *Aspergillus phoenicis* (Rizzatti et al., 2001) were purified, respectively, from the cell extract and the culture medium. Those produced by bacteria and yeasts, however, are mainly associated with the cell. Fungal β-xylosidases are often monomeric glycoproteins, but some have been reported to possess two or three subunits (Sunna and Antranikian, 1997). Generally these proteins have relatively high MWs, between 60 and 360 kDa. Although a wide range of pH optima have been observed, most lie between 4.0 and 5.0. The optimum temperature can vary from 40 to 80°C, but most β-xylosidases give best assay results at 60°C. Their thermostability is highly variable and depends on the organism in question. A good example of a stable enzyme is that from

Aspergillus phoenicis, which retained 100% of its activity after 4 h at 60°C or 21 days at room temperature (Rizzatti et al., 2001).

β-xylosidase is the main enzyme for production of monomeric xylose from solubilized xylan fragments obtained from a steaming process (Poutanen and Puls, 1988). Their action has been found to be synergistic with substituent-cleaving enzymes in the hydrolysis of substituted xylooligosaccharides. Without the presence of acetylxylan esterase, the β-xylosidase of *Trichoderma reesei* was not found to hydrolyze xylobiose bearing an acetyl substituent at the nonreducing end.

3.3 ESTERASES

Acetylxylan esterase (EC 3.1.1.6) plays an important role in the hydrolysis of xylan, as the acetyl side-groups can interfere with the approach of enzymes that break the backbone by steric hindrance, and their elimination thus facilitates the action of endoxylanases. This enzyme removes the O-acetyl groups from positions 2 and/or 3 on the β-D-xylopyranosyl residues of acetyl xylan. This enzyme was discovered late, probably because the alkaline extraction often used with highly acetylated xylans, like those in case of hardwoods, removes the acetyl groups from the xylan (Shao and Wiegel, 1992; Blum et al., 1999; Caufrier et al., 2003). Acetylxylan plays an vital role in the hydrolysis of xylan, since the acetyl side-groups can interfere with the approach of enzymes that cleave the backbone, by steric hindrance, and their elimination thus facilitates the action of endoxylanases. Few acetylxylan esterases have been purified and characterized until now and not much is known about their physicochemical properties. The major reason for the late discovery of these enzymes was the lack of a suitable substrate for their assay. As explained earlier, some xylans are acetylated in their native states, although most of the xylans used to study xylanolytic enzymes are deacetylated after alkali extraction (Tenkanen and Poutanen, 1992; Sunna and Antranikian, 1997).

Various fungi and bacteria produce acetylxylan esterases. It is important to distinguish between nonspecific acetyl esterase activity and acetylxylan esterases by using appropriate substrates. *Trichoderma reesei* culture filtrates contain two acetyl esterases, both dimeric proteins with a molecular mass of 45 kDa and pI 6.0 and 6.6. These acetyl esterases turned out to be inactive against acetylglucuronoxylan, although in the presence of such enzymes some acetate could be released from acetylated xylo-oligomers. However, in the same culture filtrates, two monomeric isoenzymatic forms (both 34 kDa) of acetyl-xylan esterases were identified as capable of liberating acetyl groups directly from polymeric xylan.

Biely (1985) first reported the presence of acetylxylan esterases in fungal cellulolytic and hemicellulolytic systems: *Trichoderma reesei*, *Aspergillus niger*, *Schizophyllum commune*, and *Aureobasidium pullulans*. Compared to plant and animal esterases, these fungal esterases exhibited high specific activities toward acetylated glucuronoxylan and were therefore named acetylxylan esterases. The production of esterases that deacetylate xylan from a number of hemicellulolytic microorganisms has been reported (Table 3.5).

Acetylxylan esterases of *T. reesei* and *A. oryzae* have been purified and characterized. The *T. reesei* enzymes had neutral isoelectric points but differed in their native molecular weights as assayed by gel chromatography. Both enzymes showed optimal activity at pH values between 5 and 6. One enzyme released only a little acetic acid from acetylated xylooligomers but showed high activity toward acetyl xylobiose. The other occurred as multiple isoenzymes and showed high activity toward acetylated xylan fragments and polymeric acetyl-4-O-methylglucuronoxylan. Enzymatic deacetylation of beechwood xylan caused precipitation of the polymer. This was shown to be due to

Table 3.5 Production of Esterases from Hemicellulolytic Microorganisms	
Butyrivibrio fibrisolvens	+
Streptomyces flavogriseus	+
Streptomyces olivochromogenes	+, 3 different
Streptomyces rubiginosus	+, 5 different
Aspergillus awamori	+
Aspergillus japonicas	+
Aspergillus niger	+
Aspergillus oryzae	+
Aspergillus versicolor	+
Fusarium oxysporum	+
Schizophyllum commune	+, 1
Trichoderma viride	+, 3 different
Trichoderma reesei	+, 2 different types purified
Based on Poutanen et al., 1991	

molecular aggregation analogous to the behavior of arabinoxylan after α-arabinosidase treatment.

3.4 ARABINASE

Arabinase hydrolyze nonreducing α-L-arabinofuranosyl groups of arabinans, arabinoxylans, and arabinogalactans. There are two types of arabinofuranosidases: Exo-acting α-L-arabinofuranosidases (EC 3.2.1.55) and Endo-1,5-α-L-arabinofuranosidases (EC 3.2.1.99). Exo-acting α-L-arabinofuranosidases are active against p-nitrophenyl-α-L-arabinofuranoside and on branched arabinans. Endo-1,5-α-L-arabinofuranosidases are active only toward linear arabinans, and are not able to hydrolyze p-nitrophenyl-α-Larabinofuranoside or arabic gum (Kaneko et al., 1993; de Vries et al., 2000). Most arabinases investigated so far are of the exo type. Arabinofuranosidases exist as monomers, but dimeric, tetrameric, and octameric forms have also been found. Reported MW values for the native enzymes range from 53 to 495 kDa, pI values range from 3.6 to 9.3, and optimum pH values from 2.5 to 6.9. All of them release arabinose from various substrates, but there is evidence of preference for particular linkages.

Most of the arabinan-degrading enzymes reported in the literature are of the exo-acting type (Polizeli et al., 2005). There are some reports of α-L-arabinofuranosidases capable of hydrolyzing both 1,3- and 1,5-α-L-arabinofuranosyl linkages in arabinoxylan. For instance, the *Aspergillus niger* enzyme first attacks the α-L-1,3-linked arabinofuranosyl residues in arabinan, and once the substrate has been hydrolyzed to approximately 30% of the original concentration, the same enzyme is then able to proceed to a slow attack of the α-L-1,5-arabinan, which eventually will be converted to arabinose (Kaji and Tagawa, 1970). Kormelink et al. (1991) reported an arabinofuranosidase able to release arabinose only from arabinoxylan. While the arabinose is released, there is no degradation of the xylan backbone as there is no production of xylooligosaccharides. This enzyme (1,4-β-D-arabinoxylan arabinofuranohydrolase) is not active against α-L-1,3- or α-L-1,5- linked arabinose from arabinans, arabinogalactans, or p-nitrophenyl-α-L-arabinofuranoside.

The production of arabinosidases in microorganisms is often associated with the production of pectinolytic or hemicellulolytic enzymes,

Table 3.6 Properties of α-arabinosidases Produced from Different Microorganisms		
Microorganism	pI	Molecular Weight (kDa)
Aspergillus niger	3.6	53
Ruminococcus albus	7.5	53
Streptomyces sp.	4.4	92
Streptomyces purpurascens	3.9	495, 62
Trichoderma reesei	6.8	305, 75
Based on Poutanen et al., 1991		

for example, in *Corticiuni rolfsii, Sclerotina fructigena, T. reesei,* and different *Streptomyces* species. Some reported molecular characteristics of α-arabinosidases are presented in Table 3.6. The purified α-arabinosidase of *Aspergillus niger,* as well as that partially purified from a commercial pectinase preparation, was able to release L-arabinose from wheat L-arabino-D-xylan. As the reaction proceeded, an amorphous precipitate consisting mainly of D-xylan with only traces of arabinose was formed. Few researchers prepared a series of arabinoxylans from purified wheat-flour arabinoxylan by partial removal of arabinosyl side branches using an α-L-arabinosidase. It was suggested that the solubilizing effect of the arabinosyl substituents was not due to increased hydration but it was due to their ability to stop intermolecular aggregation of unsubstituted xylose residues. Cereal endospermic arabinoxylans especially are known to form viscous solutions and gels. It is obvious that suitable α-arabinosidases could be used to control the degree of substitution and hence the water-binding capacity of these pentosans. In a similar way, α-galactosidases have been used in adjusting the degree of α-galactosyl substitution and hence the gelling properties of galactomannans.

3.5 α-GLUCURONIDASES

α-glucuronidases (EC 3.2.1.131) are required for hydrolysis of the α-1,2-glycosidic linkage between xylose and D-glucuronic acid or its 4-O-methyl ether. It has been found in snail digestive juice, and in several fungal and bacterial filtrates, but has been isolated to homogeneity from only few sources. The substrate specificities of α-glucuronidases differ according to enzyme source. The dimeric enzyme (MW 366 kDa, pI between 2.5 and 3.0) from *Agaricus bisporus* cleaves

4-O-methylglucuronose substituted xylo-oligomers with DP values between 2 to 6. The action of an endo-xylanase yielding substituted products of the appropriate size is a requirement to the action of the glucuronidase. Johnson et al. (1989) have reported that in contrast to *A. bisporus* enzyme, the α-glucuronidases from *Aspergillus niger* and *Schizophyllum* commune can liberate 4-O-methylglucuronic acid from 4-O-methylglucuronoxylan (Johnson et al., 1989). But the substituted fragments released by the action of endo-xylanases were found to be better substrates. Coughlan et al. (1993) reported the variation in substrate specificity among microbial α-glucuronidases. Another report described an α-glucuronidases from *Phanerochaete chrysosporium*. It showed a MW of 112 kDa and a pI value of 4.6, and optimum pH of 3.5. This enzyme showed little activity on glucuronoxylan polysaccharides but short chain xylo-oligosaccharides that were substituted with α-linked 4-O-methyl-D-glucopyranosyl uronic acid attached to the C-2 position of the nonreducing D-xylopyranosyl residue were readily hydrolyzed.

Some microorganisms exhibit their maximum activity only in the presence of short glucuronoxylan substrates. However, the substrate specificity varies with the microbial source. Some glucuronidases are able to hydrolyze the intact polymer (Puls and Schuseil, 1993; Tenkanen and Siika-aho, 2000). It has also been noted that acetyl groups close to the glucuronosyl substituents can partially hinder the α-glucuronidase activity.

The presence of acidic oligosaccharides in xylan hydrolysates produced by hemicellulolytic enzyme preparations shows the absence or insufficiency of this enzyme. 4-O-methylglucuronic acid was first found in the enzymatic hydrolysates of glucuronoxylan by Sinner et al. (1972). The presence of a uronic acid-liberating enzyme, together with β-xylosidase, was found to increase the xylose yield in the enzymatic hydrolysis of hardwood xylan (Puls et al., 1976). The presence of α-glucuronidase in the hemicellulolytic system of *T. reesei* was reported in 1983 by Dekker. The production of α-glucuronidase by many fungi and bacteria (Table 3.7) has been reported (Puls, 1992). Only a few α-glucuronidases have been totally or even partially purified and characterized (Table 3.8). The α-glucuronidase isolated from a culture filtrate of *Agaricus bisporus* by gel chromatography is a very large protein (450 kDa) (Puls et al., 1987). The enzyme has a very

Table 3.7 Production of α-glucuronidase by Different Fungi and Bacteria
A. bisporus
T. reesei
S. olivochromogenes
M. paranaguensis
S. olivochromogenes
D. dendroides
T. palustris
Trichoderma sp.
T. versicolor
L. sulphureus
A. bisporus
P. ostreatus
Streptomyces spp.
T. reesei
T. aurantiacus
Based on Puls et al., 1987; Mackenzie et al., 1987; Puls, 1992

Table 3.8 Properties of Purified Glucuronidases from Different Fungi			
Organism	pH Optimum	pI	Molecular Weight (kDa)
Agaricus bisporus	3.3–3.8	2.6–2.9	160
Trichoderma reesei Rut C-30	6.0	<4	>100
Trichoderma reesei	5.0	nd	100
Thermoascus aurantiacus	4.5	nd	118
Based on Puls, 1992; Khandke et al., 1989			

low isoelectric point and a pH optimum of about 3.3. A series of 4-O-methylglucurono-substituted xylooligosaccharides with a DP up to 5 tested as substrates showed the highest activity against 4-O-methylglucuronoxylobiose (Korte, 1980). The α-glucuronidase of *A. bisporus* had no activity toward polymeric xylan. The α-glucuronidase of *T. reesei* also had an acidic isoelectric point (Poutanen, 1988). It had a molecular weight of about 70 kDa as estimated by gel chromatography and a pH optimum at 6 with 4-O-methylglucurono xylobiose as substrate. The α-glucuronidase of the thermophilic fungus *Thermoascus aurantiacus* (Khandke et al., 1989) was a single polypeptide chain with an MW of

Table 3.9 Substrate Specificities of *T. aurantiacus* and *A. bisporus* α-Glucuronidases

Substrate	Degradation (%)	
	A. bisporus	*T. aurantiacus*
4-O-MeGlcAX	0	96
4-O-MeGlcAX2	100	100
4-O-MeGlcAX3	99	96
4-O-MeGlcAX4	96	96
4-O-MeGlcAX5	50	nd
4-O-MeGIcAX6	53	nd
4-O-MeGlcAXylan	0	52
Based on Puls, 1992		

118 kDa. The enzyme had a pH optimum at 4.5, and it hydrolyzed 4-O-methylglucurono-substituted xylooligomers from XI to X7 at rates comparable with that of xylan. The substrate specificities of *Taurantiacus* and *A. bisporus* α-glucuronidases are presented in Table 3.9 (Puls, 1992).

3.6 FERULIC ACID ESTERASE AND p-COUMARIC ACID ESTERASE

Ferulic acid esterase (EC 3.1.1.73) and p-coumaric acid esterase (EC 3.1.1.x) cleave ester bonds on xylan; the first one cleaves between arabinose and ferulic acid sidegroups, while the second one cleaves between arabinose and p-coumaric acid (Christov and Prior, 1993; Williamson et al., 1998; Crepin et al., 2004). Feruloyl esterase activity was first detected in culture filtrates of *Streptomyces olivochromogenes* (Mackenzie et al., 1987) and has thereafter also been reported for some hemicelluloyltic fungi. A partially purified feruloyl esterase from *S. commune* liberated hardly any ferulic acid without the presence of xylanase (Mackenzie and Bilous, 1988). Tenkanen et al. (1991) have purified a feruloyl esterase from *Aspergillus oryzae*. The enzyme is an acidic monomeric protein having an isoelectric point of 3.6 and a molecular weight of 30 kDa. It has wide substrate specificity, liberating ferulic, p-coumaric, and acetic acids from steam extracted wheat-straw arabinoxylan. The late discovery of acetyl xylan and feruloyl esterases has been partly due to the lack of suitable substrates. Xylans are often isolated by alkaline extraction, in which ester groups are saponified. Treatment of plant materials under mildly acidic conditions, as in steaming or aqueous-phase thermomechanical treatment, leaves most of the ester groups intact. These methods, however, partly hydrolyze xylan to shorter fragments. Polymeric acetylated xylan can be isolated from delignified materials by dimethyl sulfoxide extraction. *Streptomyces viridosporus* and several mesophilic, thermophilic, and anaerobic rumen fungi are found to produce esterases that liberate p-coumaroyl side groups (Donnelly and Crawford, 1988; Borneman et al., 1990; Smith et al., 1991).

REFERENCES

Almeida, E.M., Polizeli, M.L.T.M., Terenzi, H.F., Jorge, J.A., 1995. Purification and biochemical characterization of β-xylosidase from *Humicola grisea* var. *thermoidea*. FEMS Microbiol. Lett. 130, 171–176.

Andrade, S.V., M.L.T.M., Polizeli, Terenzi, H.F., Jorge, J.A., 2004. Effect of carbon source on the biochemical properties of the β-xylosidase produced by *Aspergillus versicolor*. Process Biochem. 39, 1931–1938.

Bajpai, P., 1997. Microbial xylanolytic enzyme system: Properties and applications. Adv. Appl. Microbiol. 43, 141–194.

Bajpai, P., 2009. Xylanases. In: Schaechter, M., Lederberg, J. (Eds.), Encyclopedia of Microbiology, Third edition. Academic Press, San Diego, pp. 600–612.

Bajpai, P., 2013. Pulp and paper bioprocessing. Encyclopedia of Industrial Biotechnology. John Wiley & Sons, Inc.

Beg, Q.K., Kapoor, M., Mahajan, L., Hoondal, G.S., 2001. Microbial xylanases and their industrialapplications: a review. Appl. Microbial. Biotechnol. 56, 326–338.

Belancic, A., Scarpa, J., Peirano, A., Diaz, R., Steiner, J., Eyzayuirre, J., 1995. *Penicillium purpurogenum* produces several xylanases: purification and properties of two of the enzymes. J. Biotechnol. 41, 71–79.

Biely, P., 1985. Microbial xylanolytic systems. Trends Biotechnol. 3, 286–290.

Biely, P., 1993. Biochemical aspects of the production of microbial hemicellulases. In: Coughlan, M.P., Hazlewood, G.P. (Eds.), Hemicelluloses and hemicellulases. Portland Press, London, pp. 29–52.

Biely, P., Markovic, O., Mislovicova, D., 1985. Sensitive detection of endo-1,4-beta-glucanases and endo-1,4-beta-xylanases in gels. Anal. Biochem. 144, 147–151.

Blum, D.L., Li, X.-L., Chen, H., Ljungdahl, L.G., 1999. Characterization of an acetyl xylan esterase from the anaerobic fungus *Orpinomyces* sp. strain PC-2. Appl. Environ. Microbiol. 65, 3990–3995.

Borneman, W.S., Hartley, R.D., Morrison, W.H., Akin, D.E., Ljungdahl, L.G., 1990. Feruloyl and p-coumaroyl esterase from anaerobic fungi in relation to plant cell wall degradation. Appl. Microbiol. Biotechnol. 33, 345–351.

Caufrier, F., Martinou, A., Dupont, C., Bouriotis, V., 2003. Carbohydrate esterase family 4 enzymes: substrate specificity. Carbohydr. Res. 338 (7), 687–692.

Chavez, R., Schachter, K., Navarro, C., Peirano, A., Aguirre, C., Bull, P., et al., 2002. Differences in expression of two endoxylanase genes (xynA and xynB) from *Penicillium purpurogenum*. Gene 293 (1–2), 161–168.

Christov, L.P., Prior, B.A., 1993. Esterases of xylan-degrading microorganisms, production, properties, and significance. Enzyme Microb. Technol. 15, 460–475.

Coughlan, M.P., Touhy, M.G., Filho, E.X.F., Puls, J., Claeyssens, M., Vrsanska, M., et al., 1993. Enzymological aspects of microbial hemicellulases with emphasis on fungal systems. In: Coughlan, M.P., Hazlewood, G.P. (Eds.), Hemicellulose and Hemicellulases. Portland Press, London, pp. 53–84.

Crepin, V.F., Fauld, C.B., Connerton, I.F., 2004. Functional classification of the microbial feruloyl esterases. Appl. Microbiol. Biotechnol. 63 (6), 647–652.

Dey, D., Hinge, J., Shendye, A., Rao, M., 1992. Purification and properties of extracellular endo-xylanases from alkalophilic thermophilic Bacillus sp. Can. J. Microbiol. 38, 436–442.

de Vries, R.P., Kester, H.C., Poulsen, C.H., Benen, J.A., Visser, J., 2000. Synergy between enzymes from Aspergillus involved in the degradation of plant cell wall polysaccharides. Carbohydr. Res. 327 (4), 401–410.

Donnelly, P.K., Crawford, D.L., 1988. Production by Streptomyces viridosporus T7A of an enzyme which cleaves aromatic acids from lignocellulose. Appl. Environ. Microbiol. 54, 2237–2244.

Elegir, G., Sykes, M., Jeffries, T.W., 1995. Differential and synergistic action of Streptomyces endoxylanases in prebleaching of kraft pulp. Enzyme Microb. Technol. 17, 954–959.

Henrissat, B., Bairoch, A., 1993. New families in the classification of glycosyl hydrolases based on amino acid sequence similarities. Biochem. J. 293, 781–788.

Johnson, D.G., Silva, M.C., Mackenzie, C.R., Schneider, H., Fontana, J.D., 1989. Microbial degradation of hemicellulosic materials. Appl. Biochem. Biotechnol. 20/21, 245–258.

Kaji, A., Tagawa, K., 1970. Purification, crystallization, and amino acid composition of α-L-arabinofuranosidase from Aspergillus niger. Biochem. Biophys. Acta 207, 456–464.

Kaneko, S., Shimasaki, T., Kusakable, I., 1993. Purification and some properties of intracellular α-L-arabinofuranosidase from Aspergillus niger 5–16. Biosci. Biotechnol. Biochem. 57, 1161–1165.

Khandke, K.M., Vithayathil, P.J., Murthy, S.K., 1989. Purification of xylanase, β-glucosidase, endocellulase, and exocellulase from a thermophilic fungus Thermoascus aurantiacus. Arch. Biochem. Biophys. 274, 511.

Kormelink, F.J.M., Searlevanleeuwen, M.I.F., Wood, T.M., Voragen, A.G.I., 1991. Purification and characterization of a (1,4)-beta-d-arabinoxylan arabinofuranohydrolase from Aspergillus awamori. Appl. Microb. Biotechnol. 35 (6), 753–758.

Korte, H.E. (1980). Ph.D. Thesis, University of Hamburg, Hamburg.

Kurakabe, M., Shinjii, O., Komaki, T., 1997. Transxylosilation of β-xylosidase from Aspergillus awamori K4. Biosci. Biotechnol. Biochem. 6112, 2010–2014.

Li, K., Azadi, P., Collins, R., Tolan, J., Kim, J.S., Eriksson, K-E L, 2000. Relationships between activities of xylanases and xylan structures. Enzyme Microb. Technol. 27, 89–94.

Mackenzie, C.R., Bilous, D., 1988. Ferulic acid esterase activity from Schizophyllum commune. Appl. Environ. Microbiol. 54, 1170.

Mackenzie, C.R., Bilous, D., Schneider, H., Johnson, K.G., 1987. Induction of cellulolytic and xylanolytic enzyme systems in Streptomyces spp. Appl. Environ. Microbiol. 53, 2835.

Polizeli, M.L., Rizzatti, A.C., Monti, R., Terenzi, H.F., Jorge, J.A., Amorim, D.S., 2005. Xylanases from fungi: properties and industrial applications. Appl. Microbiol. Biotechnol. 67, 577–591.

Poutanen K. (1988). Ph.D. Dissertation, VTT Pub. 47, Technical Research Centre of Finland.

Poutanen, K., Puls, J., 1988. Characteristics of Trichoderma reesei beta-xylosidase and its use in the hydrolysis of solubilized xylans. Appl. Microbiol. Biotechnol. 28, 425.

Poutanen, K, Tenkanen, M, Korte, H, Puls, J, 1991. Accessory enzymes involved in the hydrolysis of xylans. In: Leatham, GF, Himmel, ME (Eds.), Enzymes in Biomass Conversion. American Chemical Society; ACS Symposium series 460, Washington (DC), pp. 426–436.

Puls, J., 1992. Xylans and xylanases. In: Visser, J., Beldman, G., Kusters-van Someren, M.A., Voragen, A.G.J. (Eds.), Progress in Biotechnology, 7. Elsevier Science, Amsterdam, p. 213.

Puls J., Poutanen K. (1989). Mechanisms of enzymatic hydrolysis of hemicelluloses xylans and procedures for determination of the enzyme activities involved. In: Ericksson K.E.E., Ander P., (Eds.), Proceedings of the 3rd International Conference on Biotechnology in the Pulp and Paper Industry. Stockholm, STFI, pp. 93–95.

Puls, J., Schuseil, J., 1993. Chemistry of hemicelluloses, relationship between hemicellulose structure and enzyme required for hydrolysis. In: Coughlan, M.P., Hazlewood, G.P. (Eds.), Hemicelluloses and Hemicellulases. Portland Press, London, pp. 1–27.

Puls J., Sinner M. and Dietrichs H.H., 1976. German Pat. 2,643,800.

Puls, J., Schmidt, O., Granzow, C., 1987. α-Glucuronidase in two microbial systems. Enzyme Microb. Technol. 9, 83.

Reilly, P.J., 1981. Xylanases, structure and function. In: Hollaender, A. (Ed.), Trends in the Biology of Fermentation for Fuels and Chemicals. Plenum, New York, pp. 111–129.

Rizzatti, A.C.S., Jorge, J.A., Terenzi, H.F., Rechia, CGV, Polizeli, M.L.T.M., 2001. Purification and properties of a thermostable extracellular β-xylosidase produced by a thermotolerant Aspergillus phoenicis. J. Ind. Microbiol. Biotech. 26, 156–160.

Rizzatti, A.C.S., Sandrim, V.C., Jorge, J.A., Terenzi, H.F., Polizeli, M.L.T.M., 2004. Influence of temperature on the properties of xylanolytic enzymes of the thermotolerant fungus Aspergillus phoenicis. J. Ind. Microbiol. Biotech. 31, 88–93.

Segura, B.G., Durand, R., Fèvre, M., 1998. Multiplicity and expression of xylanases in the rumen fungus Neocallimastix frontalis. FEMS Microbiol. Lett. 164 (1), 47–53.

Shao, W., Wiegel, J., 1992. Purification and characterization of thermostable β-xylosidase from Thermoanaerobacter ethanolicus. J. Bacteriol. 17418, 5848–5853.

Sinner, M., Dietrichs, H.H., Simatupang, M.H., 1972. Holzforschung 26, 218.

Smith, D.C., Bhat, K.M., Wood, T.M., 1991. Xylan-hydrolysing enzymes from thermophilic and mesophilic fungi. World J. Microbiol. Biotechnol. 7, 475–484.

Subramaniyan, S., Prema, P., 2002. Biotechnology of microbial xylanases: Enzymology, molecular biology and application. Crit. Rev. Biotechnol. 22, 33–46.

Sung, W I., Luk, C.K., Chan, B., Wakarchuk, W., Yaguchi, M., Campbell, R., et al., 1995. Expression of Trichoderma reesei and Trichoderma viride xylanases in Escherichia coli. Biochem. Cell Biol. 73 (5–6), 253–259.

Sunna, A., Antranikian, G., 1997. Xylanolytic enzymes from fungi and bacteria. Crit. Rev. Biotechnol. 17 (1), 39–67.

Tenkanen, M., Poutanen, K., 1992. Significance of esterases in the degradation of xylans. In: Visser, J., Beldman, G., Kuster-van Someren, M.A., Voragen, A.G.J. (Eds.), Xylan and Xylanases. Elsevier, Amsterdam, pp. 203–212.

Tenkanen, M., Siika-aho, M., 2000. An alpha-glucuronidase of Schizophyllum commune acting on polymeric xylan. J. Biotechnol. 78 (2), 149–161.

Tenkanen, M., Schuseil, J., Puls, J., Poutanen, K., 1991. Production, purification, and characterization of an esterase liberating phenolic acids from lignocellulosics. J. Biotechnol. 18, 69–84.

Törrönen, A., Rouvinen, J., 1997. Structural and functional properties of low molecular weight endo-1,4-β-xylanases. J. Biotechnol. 57, 137–149.

Williamson, G., Faulds, C.B., Kroon, P.A., 1998. Specificity of ferulic acid (feruloyl) esterases. Biochem. Soc. Trans. 26 (2), 205–209.

Wong, K.K.Y., Tan, L.U.L., Saddler, J.N., Yaguchi, M., 1986a. Purification of a third distinct xylanase from the xylanolytic systems of Trichoderma harzianum. Can. J. Microbiol. 32, 570.

Wong, K.K.Y., Tan, L.U.L., Saddler, J.N., 1986b. Functional interaction among three xylanases from Trichodema harzianum. Enzyme Microb. Technol. 8, 617.

Wong, K.K.Y., Tan, L.U.L., Saddler, J.N., 1988. Multiplicity of β-1,4-xylanase in microorganisms, functions and applications. Microbiol. Rev. 52 (3), 305–317.

Zanoelo, F.F., MLTM, Polizeli, Terenzi, H.F., Jorge, J.A., 2004. Purification and biochemical properties of a thermostable xylose-tolerant β-D-xylosidase from Scytalidium thermophilum. J. Ind. Microbiol. Biotech. 31, 170–176.

CHAPTER 4

Structure and Synergism between the Enzymes of the Xylanolytic Complex

4.1 XYLANOSOMES

There have been several reports describing the presence of xylanosome in some xylan-degrading microorganisms. Xylanosomes are discrete, multifunctional, multienzyme complexes found on the surface of several microorganisms. Xylanosomes are structures analogous to the cellulosome, multiple enzyme complexes found in the surface of cells in several cellulolytic microorganisms. In addition to their degradative function, these cell-associated entities also mediate the adhesion of cells to cellulose. Cellulosomes can also be found as cell free extracellular enzyme complexes. In both forms, the cellulosome is responsible for the efficient degradation of the polymeric substrate (Bayer et al., 1994; Sunna and Antranikian, 1997).

In *Butyrivibrio fibrisolens* H17c, xylanosome exists as a multisubunit protein aggregate with a molecular weight more than 669 kDa. This contains 11 protein bands showing xylanase activity and 3 bands with endoglucanase activity (Lin and Thompson, 1991). Likewise, a multicomplex system, consisting of 7 different protein complexes, is accountable for the hydrolysis of cellulose and xylan in *Clostridium papyrosolvens* C7. The molecular weight of the complexes is found to range from 500 to 660 kDa. There is a noncatalytic glycoprotein (Subunit S4, MW 125 kDa) reported by Pohlschroder et al. (1994) that may function as a substrate binding and scaffolding component performing a similar role to that reported for the glycoprotein S1 from *Clostridium thermocellum* in all of these complexes. Jiang et al. (2005) have reported a xylanosome with a molecular weight of 1200 kDa. The analysis of the products from wheat arabinoxylan degradation by xylanosome showed that the enzyme contained endoxylanase and debranching activities. The xylotriose, xylobiose, xylose, and arabinose were found to be the major degradation products. Xylanosomes reported so far correspond to bacterial sources (Lin and Thompson, 1991; Kim and Kim, 1993; Pohlschroder et al., 1994; Deng et al., 2005; Jiang et al., 2005). *Clostridium papyrosolvens* C7 possesses a multicomplex cellulase−xylanase system, which is responsible for hydrolysis of cellulose and xylan (Pohlschroder et al., 1994). This multiplex system consists of seven protein complexes whose molecular weight ranges from 500 to 660 kDa.

4.2 SYNERGISTIC ACTION BETWEEN MULTIPLE FORMS OF XYLANASE

Multiplicity of xylanolytic enzymes has been reported in several microorganisms: *Streptomyces sp.*, *Penicillium purpurogenum*, *Melanocarpus albomyces* IIS 68, *Cellulomonas sp.* N.C.I.M 2353, and *Aeromonas caviae* W-61 (Godden et al., 1989; Belancic et al., 1995; Saraswat and Bisaria, 1997; Chaudhary and Deobagkar, 1997; Okai et al., 1998). The hydrolysis of xylan requires the action of multiple xylanases with overlapping but different specificities (Wong et al., 1988; Bajpai, 1997, 2009). The production of a multienzyme system of xylanases, in which each enzyme has a special function, is one strategy for a microorganism to achieve effective hydrolysis of xylan. During xylan hydrolysis, synergism has been observed between enzymes acting on the 1,4-β-D-xylan backbone (β-1,4-endoxylanase) and side chain-cleaving enzymes (α-L-arabinofuranosidase, acetyl xylan esterase, and β-glucuronidase). The synergistic action between acetyl xylan esterase and endoxylanases results in the efficient degradation of acetylated xylan (Biely et al., 1986). The release of acetic acid by acetyl xylan esterase increases the accessibility of the xylan backbone for endoxylanase attack. The endoxylanase creates shorter acetylated polymers, which are preferred substrates for esterase activity (Biely, 1985; Biely et al., 1985, 1986). The thermophilic actinomycete, *Thermomonospora fusca*, possesses a multienzyme system of endoxylanase, β-xylosidase, α-L-arabinofuranosidase, and acetyl esterase activities (Bachmann and McCarthy, 1991). β-Xylosidase enhances the hydrolysis of xylan by endoxylanase by relieving the end-product inhibition of endoxylanases. Similarly, the addition of α-arabinofuranosidase to endoxylanase enhances the saccharification of arabinoxylan.

Synergism between α-arabinosidase, xylanase, and β-xylosidase has been demonstrated in the hydrolysis of wheat-straw arabinoxylan with purified enzymes of *T. reesei* (Poutanen and Puls, 1989). When only xylanase and β-xylosidase were used in the hydrolysis, the xylose yield was only 66% of that produced by the whole culture filtrate at the same activity levels of these two enzymes and no arabinose was produced. Addition of α-arabinosidase increased the yields of both xylose and arabinose. Enhanced hydrolytic action of hemicellulolytic or pectinolytic enzymes in the hydrolysis of alfalfa cell wall polymers by addition of *Ruminococus albus* α-arabinosidase has also been reported (Greve et al., 1984). The synergistic action of depolymerizing and side group-cleaving enzymes has most clearly been demonstrated using acetylated xylans as substrates. Due to the high degree of acetylation, xylanases have only limited access to the polymer backbone in the absence of esterases (Poutanen and Puls, 1989). Deacetylation by acetyl xylan esterase prior to the action of xylanases, however, resulted in a lower yield than that obtained by the simultaneous action of xylanase, β-xylosidase, and esterase (Bajpai, 1997). The sequence of enzyme application not only influenced the extent of hydrolysis, but also the nature of the oligomeric end-products.

4.3 MULTIPLE FORMS OF XYLANASES

There are several reports regarding fungi and bacteria producing multiple forms of xylanases (Wong et al., 1988; Tsujibo et al., 1997). The most outstanding case regarding multiple forms of xylanases was production of more than 30 different protein bands separated by analytical electrofocusing from *Phanerochaete chrysosporium* grown in Avicel (Dobozi et al., 1992). *Streptomyces sp.* B-12-2 produces five endoxylanases when grown on oat spelt xylan (Elegir et al., 1994). The culture filtrate of *Aspergillus niger* was composed of 15, and *Trichoderma viride* of 13, xylanases (Biely et al., 1985). The filamentous fungus *Trichoderma viride* and its derivative *T. reesii* produce three cellulase free β-1,4-endoxylanases (Biely, 1985). Due to the complex structure of heteroxylans, all of the xylosidic linkages in the substrates are not equally accessible to xylan degrading enzymes because the aforementioned hydrolysis of xylan requires the action of multiple xylanases with overlapping, but different, specificities (Wong et al., 1988). The fact that protein modification (e.g., post translational cleavage) leads to the genesis of multienzymes has been confirmed by various reports (Leathers, 1988, 1989). Leathers (1988) and Li and Ljungdahl (1994) identified one xylanase, APXI, with a molecular weight of 20 kDa and later another xylanase, APX II (25 kDa), was purified by Li et al. (1993) from the same organism *Aureobasidium*. However, according to Liang et al., APXI and APXII are encoded by the gene xyn A. This suggestion was based on their almost identical N-terminal amino acid sequences, immunological characteristics, and regulatory relationships, along with the presence of a single copy of the gene and the transcript (Li and Ljungdahl, 1994). Purified APX I and APX II from *Aureobasidium pullulans* differ in their molecular weights. Post-translational modifications, such as glycosylation, proteolysis, or both, could contribute to this phenomenon (Li et al., 1993; Leathers, 1988). Therefore, several factors could be responsible for the multiplicity of xylanases. These include differential mRNA processing, post-secretional modification by proteolytic digestion, and post-translational modification, such as glycosylation and autoaggregation (Biely, 1985). Multiple xylanases can also be the product from different alleles of the same gene (Wong et al., 1988). However, some of the multiple xylanases are the result of independent genes (Hazlewood and Gilbert, 1993).

REFERENCES

Bachmann, S.L., McCarthy, A.J., 1991. Purification and cooperative activity of enzymes constituting the xylan-degrading system of *Thermomonospora fusca*. Appl. Environ. Microbiol. 57, 2121–2130.

Bajpai, P., 1997. Microbial xylanolytic enzyme system: properties and applications. Adv. Appl. Microbiol. 43, 141–194.

Bajpai, P., 2009. Xylanases. In: Schaechter, M., Lederberg, J. (Eds.), Encyclopedia of microbiology, Third edition. Academic Press, San Diego, pp. 600–612.

Bayer, E.A., Morag, E., Lamed, R., 1994. The cellulosome—a treasure trove for biotechnology. Tibtech 12, 379–386.

Belancic, A., Scarpa, J., Peirano, A., Diaz, R., Steiner, J., Eyzayuirre, J., 1995. *Penicillium purpurogenum* produces several xylanases: purification and properties of two of the enzymes. J. Biotechnol. 41, 71–79.

Biely, P., 1985. Microbial xylanolytic systems. Trends Biotechnol. 3, 286–290.

Biely, P., Markovik, O., Mislovicova, D., 1985. Sensitive detection of endo-1,4,-β-glucanases and endo-1,4-β-xylanases in gels. Anal. Biochem. 144, 147.

Biely, P., Mackenzie, C.R., Puls, J., Schneider, H., 1986. Cooperativity of esterases and xylanases in the enzymatic degradation of acetyl xylan. Biotechnology 4, 731–733.

Chaudhary, P., Deobagkar, D., 1997. Purification and characterization of xylanase from Cellulomonas sp. N.C.I.M. 2353. Biotechnol. Appl. Biochem. 25, 127–133.

Deng, W., Jiang, Z.Q., Li, L.T., Wei, Y., Shi, B., Kusakabe, I., 2005. Variation of xylanosomal subunit composition of Streptomyces olivaceoviridis by nitrogen sources. Biotechnol. Lett. 27, 429–433.

Dobozi, M.S., Szakacs, G., Bruschi, C.V., 1992. Xylanase activity of Phanerchaete chrysosporium. Appl. Environ. Microbiol. 58, 3466.

Elegir, G., Szakacs, G., Jeffries, T.W., 1994. Purification, characterization, and substrate specificities of multiple xylanases from *Streptomyces* sp. strain B-12-2. Appl. Environ. Microbiol. 60, 2609.

Godden, B., Legon, T., Helvenstein, P., Penninckx, M., 1989. Regulation of the production of hemicellulolytic and cellulolytic enzymes by a *Streptomyces* sp. growing on lignocellulose. J. Gen. Microbiol. 135, 285–292.

Greve, L.C., Labawitch, J.M., Hungate, R.E., 1984. α-L-Arabinofuranosidase from *Ruminococcus albus* 8: purification and possible role in hydrolysis of alfalfa cell wall. Appl. Environ. Microbiol. 47, 1135.

Hazlewood, G.P., Gilbert, H.J., 1993. Molecular biology of hemicellulases. In: Coughlan, M.P., Hazlewood, G.P. (Eds.), Hemicelluloses and Hemicellulases. Portland Press, London, p. 103.

Jiang, Z.Q., Deng, W., Li, X.T., Ai, Z.L., Li, L.T., Kusakabe, I., 2005. Characterization of a novel, ultra-large xylanolytic complex (xylanosome) from *Streptomyces olivaceoviridis* E-86. Enzyme Microbial. Technol. 36, 923–929.

Kim, C.H., Kim, D.S., 1993. Extracellular cellulolytic enzymes of *Bacillus circulans* are present as 2 multiple-protein complexes. Appl. Biochem. Biotechnol. 42, 83–94.

Leathers, T.D., 1988. Amino acid composition and partial sequence of xylanase from *Aureobasidium*. Biotechnol. Lett. 10, 775.

Leathers, T.D., 1989. Purification and properties of xylanase from *Aureobasidium*. J. Ind. Microbiol. 4, 341.

Li, X.L., Ljungdahl, L.G., 1994. Cloning, sequencing and regulation of a xylanase gene from the fungus *Aureobasidium pullulans* Y-2311-1. Appl. Environ. Microbiol. 60, 3160.

Li, X.L., Zhang, Z., Dean, I.F.D., Eriksson, K.L., Ljungdahl, L.G., 1993. Purification and characterization of a new xylanase (APX-II) from the fungus *Aureobasidium pullulans* Y-2311-1. Appl. Environ. Microbiol. 59, 3212.

Lin, L.L., Thompson, J.A., 1991. An analysis of the extracellular xylanases and cellulases of *Butyrivibrio fibrisolvens* H17c. Fems Microb. Lett. 84, 197–204.

Okai, N., Fukasaku, M., Kaneko, J., Tomita, T., Muramotu, K., Kamio, Y., 1998. Molecular properties and activity of a carboxyl-terminal truncated form of xylanase 3 from *Aeromonas caviae* W-61. Biosci. Biotechnol. Biochem. 62, 1560–1567.

Pohlschroder, M., Leschine, S.B., Canaleparola, E., 1994. Multicomplex cellulase xylanase system of *Clostridium papyrosolvens* C7. J. Bacteriol. 176, 70–76.

Poutanen, K., Puls, J., 1989. ACS Symp. Ser. 399, 630. The xylanolytic enzymes system of *Trichoderma reesei*. ACS Symp. Series.

Saraswat, V., Bisaria, V.S., 1997. Biosynthesis of xylanolytic and xylan-debranching enzymes in *Melanocarpus albomyces* IIS 68. J. Ferment Bioeng. 83, 352–357.

Sunna, A., Antranikian, G., 1997. Xylanolytic enzymes from fungi and bacteria. Crit. Rev Biotechnol. 17, 39–67.

Tsujibo, H., Ohtsuki, T., Iio, T., Yamazaki, I., Miyamoto, K., Sugiyama, M., et al., 1997. Cloning and sequence analysis of genes encoding xylanases and acetyl xylan esterases from *Streptomyces* OPC-520. Appl. Environ. Microbiol. 63 (2), 661.

Wong, K.K.Y., Tan, L.U.L., Saddler, J.N., 1988. Multiplicity of β-1,4-xylanase in microorganisms: functions and applications. Microbiol. Rev. 52, 305–317.

CHAPTER 5

Sources, Production, and Classification of Xylanases

5.1 SOURCES

Xylanases, the xylan hydrolyzing enzymes, are ubiquitous and diverse by nature. A number of different sources have been found to produce these enzymes, which include marine and terrestrial bacteria, rumen bacteria, fungi, marine algae, protozoa, snails, crustaceans, insects, terrestrial plants, and their seeds (Smith et al., 1991; Bajpai, 1997, 2009; Beg et al., 2001; Subramaniyan and Prema, 2002). However, filamentous fungi are particularly interesting producers of xylanases from an industrial point of view. This is because they excrete xylan-degrading enzymes into the medium, eliminating the need for cell disruption prior to purification (Sunna and Antranikian, 1997; Polizeli et al., 2005). Furthermore, xylanase levels from fungal cultures are typically much higher than those from yeast or bacteria. Also, in addition to xylanases, fungi produce several auxiliary enzymes required for the degradation of the substituted xylan (Bajpai, 2009). The genera *Aspergillus* and *Trichoderma* are pre-eminent in xylanase production among the mesophilic fungi. Efforts have been made to isolate thermophilic and even extremophilic microorganisms, as they produce enzymes of greater stability (Lasa and Berenguer, 1993; Harris et al., 1997; Ishihara et al., 1997; Kalogeris et al., 1998; Andrade et al., 1999; Niehaus et al., 1999; Puchart et al., 1999; Maheshwari et al., 2000; Rizzatti et al., 2001; Bruins et al., 2001; Monti et al., 2003; Sharma and Kumar, 2013). Several bacterial and fungal species—*Thermomonospora sp.*, *Bacillus sp.*, *Melanocarpus albomyces*, *Chaetomium thermophilum*, *Nonomuraea flexuosa*, *Streptomyces sp.*, *Dictyoglomus sp.*, *Thermotogales sp.*, *Thermoactinomyces thalophilus*, *Thermoascus aurantiacus*, *Fusarium proliferatum*, *Clostridium abusonum*, *Arthrobacter*—have been reported to produce xylanase enzymes. Noted thermophilic fungi include: *Humicola insolens*, *Humicola lanuginosa*, *Humicola grisea*, *Melanocarpus albomyces*, *Paecylomyces variotii*, *Talaromyces byssochlamydoides*, *Talaromyces emersonii*, *Thermomyces lanuginosus*, and *Thermoascus aurantiacus*, among others (Amare, 1998; Prabhu and

Maheshwari, 1999; Swaroopa and Krishna, 2000; Maheshwari et al., 2000; Kohilu et al., 2001; George et al., 2001; Hakulinen et al., 2003; Virupakshi et al., 2005; Ghatora et al., 2006; Khandeparkar and Bhosle, 2006; Garg et al., 2009, 2010). The xylanases from these fungi possess optimum temperatures between 60°C and 80°C and are very stable in this range. These enzymes are usually glycoproteins and are found to be active at an acidic pH (4.5−6.5). They exist in a multiplicity of forms, and the majority exhibit variable MWs in the range 6−38 kDa.

Endoxylanases from thermophiles have some degree of structural homology with those from mesophiles. Many researchers have attempted to explain the thermostability observed in enzymes from thermophiles in terms of extra disulphide bridges, an N-terminal proline residue causing a reduction in conformational freedom, salt bridges, and the presence of hydrophobic side-chains (Turunen et al., 2001). Hakulinen et al. (2003) also describe some minor modifications responsible for the increased thermal stability of xylanases:

− Higher Thr/Ser ratio
− Increased number of charged residues, especially Arg, resulting in enhanced polar interactions
− Improved stabilization of secondary structures involving a higher number of residues in the beta-strands, and stabilization of the alpha-helix region

According to Polizeli et al. (2005), some xylanases improve their stability by compacting the protein structure with a higher number of ion pairs or aromatic residues on the protein surface, resulting in enhanced interactions. However, no definite conclusion has been reached, as a phenomenon observed in one microorganism may not be found in another.

5.2 PRODUCTION

Xylanases may be industrially produced in submerged liquid culture or on a solid substrate. Most xylanase manufacturers produce these enzymes using submerged fermentation (SmF) techniques. In fact, SmF as a producing system accounts nearly for 90% of total xylanase produced worldwide (Polizeli et al., 2005). There is, however, a significant interest in using solid state fermentation (SSF) techniques to produce a wide variety of enzymes, including xylanases from fungal origins. SSF processes are practical for complex substrates including agricultural, forestry, and food processing residues and wastes, which are used as inducing carbon sources for the production of xylanases. SSF conditions are especially suitable for the growth of fungi as these organisms are able to grow at relatively low water activities, contrary to most bacteria and yeast, which will not proliferate under these culture conditions. Higher enzyme titres are commonly reported as an advantage for SSF processes over SmF. Additionally, enzyme properties such as thermostability and pH tolerance can also be improved when SSF is the production technique. In general, SmF is preferred as a production process when preparations require more purified enzymes, in which synergistic effects from a battery of xylan-degrading enzymes can be easily found in those preparations obtained from SSF using complex substrates. The latter is commonly sought in applications like those aimed at improving animal feed. The advantages of SSF processes over liquid batch fermentation are shown in Table 5.1.

The use of abundantly available and cost-effective agricultural residues, such as wheat bran, corn cobs, rice bran, rice husk, and other similar substrates, to achieve higher xylanase yields using SSF, allows reduction of the overall manufacturing cost of biobleached paper. This has facilitated the use of this environment-friendly technology in the paper industry. Several researchers have also shown a high yield of xylanase at various moisture levels in SSF studies. In SSF using wheat bran and eucalyptus kraft pulp as the primary solid substrates, *Streptomyces sp.* QG-11-3 (Beg et al., 2000) produced maximum xylanase yield at substrate-to-moisture ratio of 1:2.5 and 1:3, respectively. However, on increasing or decreasing the moisture level, the xylanase yield marginally decreased. In contrast, a lower solid substrate-to-moisture level of 1:1 has been reported for maximum xylanase production by *Bacillus sp.* A-009 (Gessesse and Mamo, 1999). An improvement in xylanase production by fungal mixed culture

Table 5.1 Advantages of Solid-State Fermentation over Submerged Fermentation
Smaller volumes of liquid required for product recovery,
Cheap substrate,
Low cultivation cost for fermentation,
Lower risk of contamination.

(*Trichoderma reesei* LM-UC4 E 1, *Aspergillus niger* ATCC 10864, and *A. phoenicis* QM 329) using SSF has also been reported (Gutierrez-Correa and Tengerdy, 1998). A higher xylanase yield using SSF compared with submerged fermentation using wheat straw and sugarcane bagasse has been reported from thermophilic *Melanocarpus albomyces* IIS-68 (Jain, 1995).

In cultures on solid substrate, wheat bran and rice are regarded as inducers. In cultures on solid substrate, wheat bran and rice induce xylanase production (Kadowaki et al., 1995; Damaso et al., 2000; Medeiros et al., 2000; Pandey et al., 2000; Singh et al., 2000; Anthony et al., 2003). In liquid culture, xylanase is produced in response to xylans from various sources (Gomes et al., 1994; Liu et al., 1999; Rani and Nand, 2000). β-D-Xylopyranosyl residues can also act as inducers of the xylanolytic complex (Ghosh and Nanda, 1994; Rizzatti et al., 2001), but in some microorganisms this can give rise to discriminatory control, leading to catabolite repression of endoxylanases (Flores et al., 1996; Mach et al., 1996). Another compound often used as a potent inducer is β-methyl xyloside, a nonmetabolizable structural analogue of xylobiose that can be made at low cost (Morosoli et al., 1987; Simão et al., 1997a, b). Induction of the xylanolytic system by other synthetic compounds, such as 2-O-β-D-xylopyranosyl D-xylose (Xylβ-2Xyl), 3-O-β-D-xylopyranosyl D-xylose (Xylβ1-3Xyl), and 2-O-β-D-glucopyranosil D-xylose (Glcβ1-2Xyl), has also been described (Hrmová et al., 1991). The xylobioses, which are homodisaccharides (Xylβ1-2Xyl e Xylβ1- 3Xyl), are potent inducers of endo-1,4-β-xylanase but fail to induce an enzyme of the cellulolytic complex, endo-1,4-β-glucanase. The opposite is true for the heterosaccharide Glcβ1-2Xyl. Such synthetic substrates serve as hybrid inducers, promoting the synthesis of both of the enzyme complexes (Polizeli et al., 2005).

Another way to increase xylanase production and to reduce the cost of the enzyme is by the isolation of overproducing mutants, as most of the studies have been conducted with the wild type strains. Singh et al. (1995) have isolated a mutant of *Fusarium oxysporum*. This mutant showed high activity of both xylanase and β-xylosidase of up to three-fold in comparison to the parental strain. This mutant produced high levels of xylanolytic activity on commercial xylan and also on several agricultural residues, of which wheat bran produced maximum enzyme yields. Parasexual recombination between overproducing strains has also proved to be a convenient method to improve xylanase production by *Aspergillus* strains (Loera and Córdova, 2003). Also, recombinant DNA technology has been successfully employed and will become important to overproduce xylanases in different host organisms. Yet, most xylanase preparations are still obtained from naturally overproducing microorganisms. Commercial preparation of xylanases is limited to a great extent to *Trichoderma* spp. and *Aspergillus* spp. But this might change in the future, since several promising microorganisms have been described as xylanase producers. These enzymes show increased activity and other desirable properties, for example, thermostability, stability under acid or alkaline conditions, or lack of cellulase activity. Significant progress has been made in identifying process parameters that produce higher levels of xylanase and influence the economics of the xylanase production process. Commercial xylanase preparations are manufactured by several companies in the world as shown in Table 1.1 of Chapter 1.

5.3 CLASSIFICATION OF XYLANASES

Endoxylanases are found in families 5, 7, 8, 10, 11, 26, and 43. Endoxylanases from families 10 (formerly known as F) and 11 (formerly known as G) have been described in literature. Family F/10 endo-xylanases are high molecular weight enzymes structurally composed of a cellulose-binding domain and a catalytic domain connected by a linker peptide. Xylanases belonging to this family have a (β/α) 8 fold TIM barrel (Biely et al., 1997). Family G/11 endoxylanases contain low molecular weight enzymes. Based on their isoelectric points, family 11 xylanases are further subgrouped into two: alkaline and acidic pIs. It is considered that due to their relatively small sizes these endoxylanases can pass through the pores of hemicellulose networks bringing efficient hydrolysis. These enzymes have a β-jelly roll structure (Törrönen and Rouvinen, 1997).

Wong et al. (1988) have classified microbial xylanases into two groups on the basis of their physicochemical properties, such as molecular mass and isoelectric point, rather than on their different catalytic properties. While one group consists of high molecular mass enzymes with low pI values and the other of low molecular mass enzymes with high pI values, there are exceptions. The aforementioned observation was later found to be in tune with the classification of glycanases on the basis of hydrophobic cluster analysis and sequence similarities (Sapag et al., 2002). As already discussed, the high molecular weight endoxylanases with low pI values belong to glycanase family 10 while the low molecular mass endoxylanases with high pI values are classified as glycanase family 11 (Kuno et al., 2000). There has been the addition of 123 proteins in family 11, out of which 113 are xylanases/ ORFs for xylanases, 1 is an unnamed protein, and 9 are sequences from US patent collection. But, 150 members are present in family 10, of which 112 have xylanase activities. Biely et al. (1985), after extensive study on the differences in catalytic properties among the xylanase families, concluded that endoxylanases of family 10—in contrast to the members of family 11—are capable of attacking the glycosidic linkages next to the branch points and toward the nonreducing end (Biely et al., 1997). While endoxylanases of family 10 require 2 unsubstituted xylopyranosyl residues between the branches, endoxylanases of family 11 require 3 unsubstituted consecutive xylopyranosyl residues. According to Biely et al. (1997), endoxylanases of family 10 possess several catalytic activities, which are compatible with β-xylosidases. The endoxylanases of family 10 liberate terminal xylopyranosyl residues attached to a substituted xylopyranosyl residue, but they also exhibit aryl-β-D-xylosidase activity. After conducting an extensive factor analysis study, Sapag et al. (2002) applied a new method without referring to previous sequence analysis for classifying family 11 xylanases, which could be subdivided into six main groups. Groups I, II, and III contain mainly fungal enzymes. The enzymes in groups I and II are generally 20 kDa enzymes from *Ascomyceta* and *Basidiomyceta*. The group I enzymes have basic pI values, while those of group II exhibit acidic pIs. Enzymes of group III are mainly produced by anaerobic fungi. In contrast, the bacterial xylanases are divided into three groups (A, B, and C). Group A mainly contains enzymes produced by members of the *Actinomycetaceae* and the *Bacillaceae* families; they are strictly aerobic gram-positive. Groups B and C are more closely related and mainly contain enzymes from anaerobic gram-positive bacteria, which usually live in the rumen. Xylanases from aerobic gram-negative bacteria are found in subgroup Ic, as they closely resemble the fungal enzymes of group I. Unlike previous classifications, a fourth group of fungal xylanases consisting of only two enzymes has been reported (Sapag et al., 2002).

β-Xylosidases are grouped into five families of glycoside hydrolases: 3, 39, 43, 52, and 54. Enzymes from families 3, 39, 52, and 54 catalyze hydrolysis of xylooligomers via the retaining mechanism, while enzymes from family 43 perform hydrolysis by inverting the anomeric configuration. According to the CAZY database (www. cazy.org), currently the nucleotide entries, protein data, and crystal structures are only available for families 3, 39, 52, and 54 β-xylosidases (Yang et al., 2004).

REFERENCES

Amare, G., 1998. Purification and properties of two thermostable alkaline xylanases from an alkaliphilic *Bacillus sp*. Appl. Environ. Microbiol. 64, 3533–3535.

Andrade, C.M.M.C., Pereira Jr., N., Antranikian, G., 1999. Extremely thermophilic microorganisms and their polymer-hydrolytic enzymes. Rev. Microbiol. 30, 287–298.

Anthony, T., Raj, K.C., Rajendran, A., Gunasekaran, P., 2003. High molecular weight cellulase-free xylanases from alkali-tolerant *Aspergillus fumigatus* AR1. Enzyme. Microb. Technol. 32, 647–654.

Bajpai, P., 1997. Microbial xylanolytic enzyme system: properties and applications. Adv. Appl. Microbiol. 43, 141–194.

Bajpai, P., 2009. Xylanases. In: Schaechter, M., Lederberg, J. (Eds.), Encyclopedia of Microbiology, Third edition. Academic Press, San Diego, pp. 600–612.

Beg, Q.K., Bhushan, B., Kapoor, M., Hoondal, G.S., 2000. Enhanced production of a thermostable xylanase from Streptomyces sp. QG-11-3 and its application in biobleaching of eucalyptus kraft pulp. Enzyme. Microb. Technol. 27, 459–466.

Beg, Q.K., Kapoor, M., Mahajan, L., Hoondal, G.S., 2001. Microbial xylanases and their industrial applications: a review. Appl. Microbiol. Biotechnol. 56, 326–338.

Biely, P., Markovic, O., Mislovicova, D., 1985. Sensitive detection of endo-1,4-beta-glucanases and endo-1,4-beta-xylanases in gels. Anal. Biochem. 144, 147–151.

Biely, P., Vrsanská, M., Tenkanen, M., Kluepfel, D., 1997. Endo-α-1,4- xylanase families: differences in catalytic properties. J. Biotechnol. 57, 151–166.

Bruins, M.E., Janssen, A.E., Boom, R.M., 2001. Thermozymes and their applications: a review of recent literature and patents. Appl. Biochem. Biotechnol. 90, 155–186.

Damaso, M.C.T., Andrade, C.M.M.C., Pereira Jr., N., 2000. Use of corncob for endoxylanase production by thermophilic fungus *Thermomyces lanuginosus* IOC-4145. Appl. Biochem. Biotechnol. 84–86, 821–834.

Flores, M.E., Perea, M., Rodríguez, O., Malváez, A., Huitron, C., 1996. Physiological studies on induction and catabolic repression of β-xylosidase and endoxylanase in *Streptomyces* sp. CH-M- 1035. J. Biotechnol. 49, 179–187.

Garg, N., Mahatman, K.K., Kumar, A., 2010. Xylanase: Applications and Biotechnological Aspects. Lambert Academic Publishing AG & Co., KG (Germany).

Garg, S., Ali, R., Kumar, A., 2009. Production of alkaline xylanase by an alkalo-thermophilic bacteria *Bacillus halodurans* MTCC 9512 isolated from dung. Curr. Trends Biotech. Pharm. 3, 90–96.

George, S.P., Ahmad, A., Rao, M.B., 2001. A novel thermostable xylanase from *Thermomonospora* sp.: influence of additives on thermostability. Bioresour. Technol. 78, 221–224.

Gessesse, A., Mamo, G., 1999. High-level xylanase production by an alkalophilic *Bacillus* sp. by using solid-state fermentation. Enzyme. Microb. Technol. 25, 68–72.

Ghatora, S.K., Chadha, B.S., Badhan, A.K., Saini, H.S., Bhat, M.K., 2006. Identification and characterization of diverse xylanases from thermophilic and thermotolerant fungi. BioResources 1, 18–33.

Ghosh, M., Nanda, G., 1994. Physiological studies on xylose induction and glucose repression of xylanolytic enzymes in *Aspergillus sydowii* MG49. FEBS Microbiol. Lett. 117, 151–156.

Gomes, D.J., Gomes, J., Steiner, W., 1994. Factors influencing the induction of endo-xylanase by *Thermoascus aurantiacus*. J. Biotechnol. 33, 87–94.

Gutierrez-Correa, M., Tengerdy, R.P., 1998. Xylanase production by fungal mixed culture solid substrate fermentation on sugarcane bagasse. Biotechnol. Lett. 20, 45–47.

Hakulinen, N., Turunen, O., Janis, J., Leisola, M., Rouvinen, J., 2003. Three-dimensional structures of thermophilic beta-1,4-xylanases from *Chaetomium thermophilum* and *Nonomuraea flexuosa*. Comparison of twelve xylanases in relation to their thermal stability. Eur. J. Biochem. 270, 1399–1412.

Harris, G.W., Pickersgill, R.W., Connerton, I., Debeire, P., Touzel, J.-P., Breton, C., et al., 1997. Structural basis of the properties of an industrially relevant thermophilic xylanase. Proteins 29, 77–86.

Hrmová, M., Petraková, E., Biely, P., 1991. Induction of cellulose- and xylan-degrading enzyme system in Aspergillus terreus by homo- and heterodisaccharides composed of glucose and xylose. J. Gen. Microbiol. 137, 541–547.

Ishihara, M., Tawata, S., Toyama, S., 1997. Purification and some properties of a thermostable xylanase from thermophilic fungus strain HG-1. J. Ferment. Bioeng. 835, 478–480.

Jain, A., 1995. Production of xylanase by thermophilic Melanocarpus albomyces IIS 68. Process Biochem. 30, 705–709.

Kadowaki, M.K., Pacheco, M.A.C., Peralta, R.M., 1995. Xylanase production by Aspergillus isolates grown on corn cob. Rev. Microbiol. 263, 219–223.

Kalogeris, E., Christakopoulos, P., Kekos, D., Macris, B.J., 1998. Studies on solid-state production of thermostable endoxylanases from *Thermoascus aurantiacus*. Characterization of two isozymes. J. Biotechnol. 60, 155–163.

Khandeparkar, R.D.S., Bhosle, N.B., 2006. Isolation, purification and characterization of the xylanase produced by Arthrobacter sp. MTCC 5214 when grown in solid-state fermentation. Enzyme. Microbial. Technol. 39, 732–742.

Kohilu, U., Nigam, P., Singh, D., Chaudhary, K., 2001. Thermostable, alkaliphilic and cellulose free xylanases production by *Thermoactinomyces thalophilus* subgroup C. Enzyme. Microb. Technol. 28, 606–610.

Kuno, A., Kaneko, S., Ohtsuki, H., Ito, S., Fujimoto, Z., Mizuno, H., et al., 2000. Novel sugar-binding specificity of the type XIII xylan-binding domain of a family F/10 xylanase from *Streptomyces olivaceoviridis* E-86. FEBS Lett. 482, 231.

Lasa, I., Berenguer, J., 1993. Thermophilic enzymes and their biotechnological potential. Microbiology 9, 77–89.

Liu, W., Lu, Y., Ma, G., 1999. Induction and glucose repression of endo-β-xylanase in the yeast *Trichosporon cutaneum* SL409. Process Biochem. 34, 67–72.

Loera, O., Córdova, J., 2003. Improvement of xylanase production by a parasexual cross between *Aspergillus niger* strains. Braz. Arch. Biol. Technol. 46, 177–181.

Mach, R.L., Strauss, J., Zeilinger, S., Schindler, M., Kubicek, C.P., 1996. Carbon catabolite repression of xylanase I xyn1 gene expression in *Trichoderma reesei*. Mol. Microbiol. 216, 1273–1281.

Maheshwari, R., Bharadwaj, G., Bhat, M.K., 2000. Thermophilic fungi: their physiology and enzymes. Microbiol. Mol. Biol. Rev. 64, 461–488.

Medeiros, R.G., Soffner, M.A.P., Tomé, J.A., Cacais, A.O.G., Estelles, R.S., Salles, B.C., et al., 2000. The production of hemicellulases by aerobic fungi on medium containing residues of banana plant as substrate. Biotechnol. Prog. 16, 522–524.

Monti, R., Cardello, L., Custódio, M.F., Goulart, A.J., Sayama, A.H., Contiero, J., 2003. Production and purification of an endo-1,4-β-xylanase from *Humicola grisea* var. *thermoidea* by electroelution. Braz. J. Microbiol. 34, 124–128.

Morosoli, R., Durand, S., Letendre, E., 1987. Induction of xylanase by β-methylxyloside in *Cryptococcus albidus*. FEMS. Microbiol. Lett. 48, 261–266.

Niehaus, F., Bertoldo, C., Kähler, M., Antranikian, G., 1999. Extremophiles as a source of novel enzymes for industrial application. Appl. Microbiol. Biotechnol. 51, 711–729.

Pandey, A., Soccol, C.R., Nigam, P., Soccol, V.T., 2000. Biotechnological potential of agro-industrial residues. I, sugarcane bagasse. Bioresour. Technol. 74, 69–80.

Polizeli, M.L., Rizzatti, A.C., Monti, R., Terenzi, H.F., Jorge, J.A., Amorim, D.S., 2005. Xylanases from fungi: properties and industrial applications. Appl. Microbiol. Biotechnol. 67, 577–591.

Prabhu, K.A., Maheshwari, R., 1999. Biochemical properties of xylanases from a thermophilic fungus, *Melanocarpus albomyces*, and their action on plant cell walls. J. Biosci. 4, 461–470.

Puchart, V., Katapodis, P., Biely, P., Kremnický, L., Christakopoulos, P., Vršanská, M., et al., 1999. Production of xylanases, mannanases and pectinases by the thermophilic fungus *Thermomyces lanuginosus*. Enzyme. Microb. Technol. 24, 355–361.

Rani, D.S., Nand, K., 2000. Production of thermostable cellulase-free xylanase by *Clostridium absonum* CFR-702. Process Biochem. 36, 355–362.

Rizzatti, A.C.S., Jorge, J.A., Terenzi, H.F., Rechia, C.G.V., Polizeli, M.L.T.M., 2001. Purification and properties of a thermostable extracellular β-xylosidase produced by a thermotolerant *Aspergillus phoenicis*. J. Ind. Microbiol. Biotech. 26, 156–160.

Sapag, A., Wouters, J., Lambert, C., de Ioannes, P., Eyzaguirre, J., Depiereux, E., 2002. The endoxylanases from family 11: computer analysis of protein sequences reveals important structural and phylogenetic relationships. J. Biotechnol. 95, 109–131.

Sharma, M., Kumar, A., 2013. Xylanases: an overview. Br. Biotechnol. J. 3 (1), 1–28.

Simão, R.C.G., Souza, C.G.M., Peralta, R.M., 1997a. The use of methyl β-D-xyloside as a substrate for xylanase production by *Aspergillus tamarii*. Can. J. Microbiol. 43, 56–60.

Simão, R.C.G., Souza, C.G.M., Peralta, R.M., 1997b. Induction of xylanase in *Aspergillus tamarii* by methyl β-D-xyloside. Appl. Microbiol. Biotechnol. 47, 267–271.

Singh, A., Kuhad, R.C., Kumar, M., 1995. Xylanase production by a hyperxylanolytic mutant of *Fusarium oxysporum*. Enzyme. Microb. Technol. 17, 551–553.

Singh, S., Reddy, P., Haarhoff, J., Biely, P., Janse, B., Pillay, B., et al., 2000. Relatedness of *Thermomyces lanuginosus* strains producing a thermostable xylanase. J. Biotechnol. 81, 119–128.

Smith, D.C., Bhat, K.M., Wood, T.M., 1991. Xylan-hydrolysing enzymes from thermophilic and mesophilic fungi. World J. Microbiol. Biotechnol. 7, 475–484.

Subramaniyan, S., Prema, P., 2002. Biotechnology of microbial xylanases: enzymology, molecular biology and application. Crit. Rev. Biotechnol. 22, 33–46.

Sunna, A., Antranikian, G., 1997. Xylanolytic enzymes from fungi and bacteria. Crit. Rev. Biotechnol. 17, 39—67.

Swaroopa, R.D., Krishna, N., 2000. Production of thermostable cellulase-free xylanase by Clostridium absonum CFR-702. Process Biochem. 36, 355—362.

Törrönen, A., Rouvinen, J., 1997. Structural and functional properties of low molecular weight endo-1,4-β-xylanases. J. Biotechnol. 57, 137—149.

Turunen, O., Etuaho, K., Fenel, F., Vehmaanperä, J., Wu, X., Rouvinen, J., et al., 2001. A combination of weakly stabilizing mutations with a disulfide bridge in the α-helix region of *Trichoderma reesei* endo-1,4-β-xylanase II increases the thermal stability through synergism. J. Biotechnol. 88, 37—46.

Virupakshi, K., Kyu, K.L., Tanticharoen, M., 2005. Purification and properties of a xylan-binding endoxylanase from alkaliphilic *Bacillus* sp. strain K-1. Appl. Environ. Microbiol. 65, 694—697.

Wong, K.K.Y., Tan, L.U.L., Saddler, J.N., 1988. Multiplicity of β-1,4-xylanase in microorganisms, functions and applications. Microbiol Rev. 52 (3), 305—317.

Yang, J.K., Yoon, H.-J., Ahn, H.J., Il Lee, B., Pedelacq, J.D., Liong, E.C., 2004. Crystal structure of β-D-xylosidase from *Thermoanaerobacterium saccharolyticum*, a family 39 glycoside hydrolase. J. Mol. Biol. 335, 155—165.

CHAPTER 6

Purification of Xylanases

Enzyme purification is of great importance in to acquire knowledge about structural and functional properties and to foretell its applications. The ultimate degree of purity of a particular enzyme depends upon its end use. The objective behind deciding the strategy for purification is to obtain the greatest possible yield of the desired enzyme with the highest catalytic activity and the greatest possible purity. Most of the purification methods used in laboratory research can be easily scaled to industrial processes. These methods are:

- Filtration
- Centrifugation
- Ultrafiltration
- Diafiltration
- Precipitation
- Chromatography, such as ion-exchange chromatography and gel filtration chromatography

The key for successful and efficient purification strategies is the selection of the appropriate techniques that maximize yield and purity with the minimum number of steps (Palmer, 2001).

Xylanases are mostly found to be extracellular, but intracellular xylanases have also been reported in some organisms, such as *Ruminicoccus flavefaciens* (Fontes et al., 2000), *Bacillus stearothermophilus* T-6 (Teplitsky et al., 2000), *Penicillium janthinellum* (Rodrigues and Tambourgi, 2001), and others.

Purification of xylanases to homogeneity is required for complete biochemical and molecular studies and for the determination of their primary amino-acid sequences and their three dimensional structures (Sa-Pereira et al., 2003). Methods for protein purification vary from a simple one-step purification procedure to large-scale purification processes.

Purification of xylanases is usually based on the multistep series of nonspecific techniques (Wong and Saddler, 1992; Subramaniyan and

Prema, 2002). Xylanase purification schemes have generally used standard column chromatographic techniques, mainly ion exchange and gel filtration, but also hydrophobic interaction. The low molecular weight of certain xylanases has also enabled their separation from other proteins using ultrafiltration. Furthermore, adsorptive interactions of certain low-molecular-weight xylanases (20 kDa) with gel-filtration resin matrices cause these enzymes to elute as proteins smaller than 12 kDa. This phenomenon has probably facilitated xylanase purification in certain cases, and it is probably due to the affinity of dextran gel matrices for the aromatic group of certain acids. The various purification schemes have successfully isolated a number of apparent xylanases from *Trichoderma sp.* to electrophoretic homogeneity. There is, however, some evidence that one of the low-molecular-weight xylanases from *T. viride* can be nonhomogenous under denaturing conditions, showing smaller peptides under SDS-PAGE or urea-SDS-PAGE. These fragments may be due to the cleavage of certain peptide linkages by proteases excreted by the fungus. For application on pulp fibers, the purification of xylanases does not have to be complete as long as cellulase activity is eliminated. Various strategies for preparing cellulase-free xylanases have been reviewed. Ultrafiltration can be used in conjunction with solvent exchange through ion exchangers to produce high yields of cellulase-free xylanase from the culture filtrate of *T. harzianum*. Alternatively, selective adsorption of cellulases to a cellulose column may be used to enrich xylanases; however, the column would also retain certain xylanases that contain cellulose-binding domains (Bajpai, 2009).

One of the most important steps in purification of extracellular xylanases is clarification. This step is mainly used to remove polyphenols, pigments, and nucleic acids, among others, which may be contaminants during further purification steps. Clarification can be done by precipitation, centrifugation, or filtration. This method is very rarely mentioned in the literature for xylanases, but there are certain reports that have utilized it for purification of xylanase. Dhillon et al. (2000) and Nakamura et al. (1993) reported clarification of xylanase by centrifugation at high and moderate speed, respectively.

Microbial xylanases are mainly purified by chromatographic methods, using from two to five purification steps and providing recovery yields ranging from 0.2 to 78%. As expected, lower yields were generally obtained when a greater number of purification steps are used.

A fewer number of steps generally have higher recovery yields. Precipitation with ammonium sulfate and ethanol was used in the maximum purification studies as the initial step. It also removes low molecular proteins (Nakamura et al., 1993). An alternative to precipitation is ultrafiltration. The membranes are used with a 5−30 kDa molecular-weight cut-off value (Sa-Pereira et al., 2003). A large number of purification protocols for xylanases mention ammonium sulphate precipitation as the preliminary step in purification. Ninawe et al. (2008) reported 1.4-fold purity of xylanase from *Streptomyces cyaneus* SN32 after 60% saturation with ammonium sulphate. Chapla et al. (2011) also reported 13.09-fold purification of xylanase from *Paenibacillus sp.* ASCD2 after ammonium sulphate precipitation followed by dialysis.

Gupta et al. (1994) purified xylanase from *Trichoderma viride* by precipitation with an ionic polymer Eudragit S 100. The purified enzyme was tested free from contaminant proteins and recovery was 89%. They showed that binding of the enzyme to the polymer was predominantly by electrostatic interaction. Bronnenmeier et al. (1996) purified cellulose-binding enzymes including xylanase from *Clostridium stercorarium* using cellulose-affinity chromatography. Santos et al. (2002) purified xylanase to near homogeneity using packed and expanded beds and by making use of cationic adsorbent, Streamline SP, in the presence of *Bacillus pumilus* cell mass. Breccia et al. (1999) isolated extracellular xylanase from the whole broth of *Bacillus amyloliquefaciens* by adsorption on a cation exchanger, Amberlite IRC-50, in a fluidized bed with a low degree of expansion. They eluted the adsorped enzyme from Amberlite IRC-50 by an increase in pH. The enzyme was sufficiently pure with 82% yield. Higher yield with sufficient purity with bioseparation is advantageous for industrial applications.

Dhillon et al. (2000) used sepharose Q anion exchange chromatography for purification of xylanase from *Bacillus circulans* AB16. DEAE sepharose anion exchange chromatography was used for purification of xylanase from *Trichoderma reesei* (Lappalainen et al., 2000). Use of cation exchange chromatography is also reported using carboxymethyl sepharose for purification of xylanase from *Fibrobacter succinogens* S85 and *Streptomyces rosei* (Matte and Forsberg, 1992; Lappalainen et al., 2000). Carmona et al. (2005) and Fengxia et al. (2008) reported the use of DEAE sephadex for the purification of xylanase from *Aspergillus versicolor* and *Aspergillus ficuum* AF-98, respectively. DEAE cellulose column chromatography was also used for the purification of xylanase and color removal from the extract of alkali-tolerant *Aspergillus fischeri* Fxn1 (Chandra and Chandra, 1996). Zhiwei et al. (2008) obtained 11.6-fold purification of xylanase after DEAE cellulose column chromatography from microbial community EMSD5. Gel permeation chromatography was the second-most employed purification method after ion exchange chromatography. Milagres et al. (2005) reported the use of gel permeation chromatography with Sephadex G-50 as the matrix for purification of xylanase from *Ceriporiopsis subvermispora*. Mamo et al. (2006) used two different gel permeation chromatographies for purification of endoxylanase from *Bacillus halodurans* S7. Two xylanases were purified from *Aspergillus caespitosus* using gel permeation chromatography with a matrix of Sephadex G-100 (Sandrim et al., 2004). Single endoxylanase was also purified from *Paenibacillus sp.* ASCD2 using single-step purification by gel permeation chromatography (Chapla et al., 2011). In maximum cases, the gel filtration chromatography has been used in combination with other chromatography for purification of xylanase, however single use of gel permeation chromatography is also reported (Lee et al., 1993).

Apart from ion exchange and gel permeation chromatography, other techniques—such as affinity chromatography, hydroxyapatite chromatography, and HPLC, among others—have been done in special cases. As such, classical affinity chromatography has not proven to be amenable to large scale work due to its high expense and the short life of the matrices. However, a few reports are available that used such techniques to purify xylanase. Lee et al. (1993) reported the use of affinity chromatography for purification of xylanase from *Thermoanaerobacterium saccharolyticum* B6A-RI. The chromatographic methods are sometimes coupled with HPLC columns for better resolution than the standard conventional column. Rao et al. (1996) reported the use of this type of complex system for purification of xylanase from *Chainia sp.* and used a reverse-phase HPLC column. A minor form of xylanase from *Aspergillus versicolor* has been purified using a combination of various chromatographies and finally eluted from the HPLC GF-510 system for complete purification (Carmona et al., 2005). With the advancement in instrumentation facilities, another chromatography—such as fast protein liquid chromatography—has emerged as an efficient tool for protein purification. Jorgensen

Table 6.1 Purification of *Bacillus* sp. GRE7 Xylanase			
Purification	Total Protein (mg)	Specific Activity (IU mg^{-1})	Purification (fold)
Culture supernatant	680	48.8	1.0
Ammonium sulphate precipitation	124	191.1	3.9
DEAE-cellulose	27.4	582.9	11.9
Sephadex G-75	6.4	1,392.6	28.5
Based on Kiddinamoorthy et al., 2008			

et al. (2003) used ion exchange and hydrophobic interaction chromatography on an FPLC system for the purification of xylanase from *Penicillium brasilianum* IBT20888. Li et al. (2005) also characterized cellulase-free, neutral xylanase from *Thermomyces lanuginosus* CBS 288.54, purified with the help of a Q-Sepharose fast flow column. Table 6.1 shows the biochemical properties of purified xylanase from different sources (Sa-Pereira et al., 2003).

Roy and Habib (2009) collected bacteria capable of producing clear xylanolytic and transparent zones on the xylan agar plate and incubated at 37°C for 48 h. They used pure culture technique for the collection, followed by the staining method, to screen the colonies and analyze those colonies in the microscope. They found 94% identity of these strains with *Bacillus cereus* on sequencing of the 16S rRNA gene. After isolation and screening of these bacteria, they purified the xylanase in four steps of a purification process using ammonium sulphate precipitation, DEAE-sepharose, Phenyl-5PW, and Hydroxyapatite column chromatography. With a yield of 13.4%, they purified the xylanase up to 16.3-fold. A single band of 31 kDa was determined after analysis of the purified enzyme by SDS-PAGE.

Purification of the Xylanase enzymes from *Penicillium janthinellum* and *Neurospora crassa* have been done with ammonium sulfate precipitation followed by dialysis (Abirami et al., 2011; Kumar et al., 2011; Irfan and Syed, 2012; Kamble and Jadhav, 2012). After dialysis, dialyzed enzyme was loaded onto DEAE-cellulose ion-exchange column chromatography. Compared to *Neurospora crassa*, *Penicillium janthinellum* was found to be a better xylanase producer. Abirami et al. (2011) and Irfan and Syed (2012) also used the same method for purification of a xylanase enzyme produced by the solid-state fermentation process using *Trichoderma viride*-IR05.

Yasinok et al. (2010) reported the 186-fold purification of xylanase from *Bacillus pumilus* SB-M13A by hydrophobic interaction chromatography. The xylanase was stable at alkaline pH and maximum activity was observed at 60°C and pH 7.5. In an aqueous-two-phase system, xylanase always partitioned to the top phase. Basic pH, low PEG concentration, salt addition, and the presence of microbial cells enhanced xylanase partitioning. A maximum 7-fold purification, 10-fold concentration, and 100% xylanase recovery were obtained, separately, by adjusting system parameters. The xylanase gene was amplified using PCR and a remarkable sequence similarity to xylanase genes of other *Bacillus pumilus* strains was observed. The enzyme had high substrate affinity and catalytic activity.

A highly thermostable alkaline xylanase was purified to homogeneity from a culture supernatant of *Bacillus sp.* JB 99 using DEAE-Sepharose and Sephadex G-100 gel filtration with a 25.7-fold increase in activity and a 43.5% recovery. The molecular weight of the purified xylanase was found to be 20 kDa by SDS-PAGE and zymogram analysis. The enzyme was optimally active at 70°C, pH 8.0, and was stable over a pH range of 6.0–10.0. The relative activity at 9.0 and 10.0 was 90% and 85% of that of pH 8.0, respectively. The enzyme showed high thermal stability at 60°C, with 95% of its activity after 5 hours. The Km and Vmax of the enzyme for oat spelt xylan were 4.8 mg/ml and 218.6 µM/minute/mg, respectively. Analysis of the N-terminal amino acid sequence revealed that the xylanase belonged to glycosyl hydrolase family 11, from thermoalkalophilic *Bacillus sp.*, with a basic pI (Shrinivas et al., 2010).

Kiddinamoorthy et al. (2008) studied purification of *Bacillus sp.* GRE7 xylanase. Ammonium sulphate (40–80%) precipitation resulted in a 71% recovery of initial xylanase activity in the culture supernatant. Fractions eluted during ion-exchange (DEAEcellulose) chromatography gave two active xylanase components. Multiplicity of high- and low-molecular weight xylanases observed in *Bacillus spp.* was explained to be due to distinct coding genes (Wong et al., 1988), differential glycosylation at post-translation, or proteolysis. The major and minor components constituted about 48% and <3% of the total xylanase activity, respectively, but only the major component was purified further by gel filtration (Sephadex G-75). The latter procedure yielded an overall xylanase recovery of 27% and 28.5-fold purification. SDS-PAGE revealed a single band,

Table 6.2 Properties of Purified Xylanase from Different Microorganisms

Microorganisms	No. of Steps	Yield (%)	Purification Factor (Fold)	Molecular Weight (kDa)	Specific Activity (U/mg)
Aspergillus niger	1	70	65	24	31.4
Thermotoga maritima MSB8	4	31	4.4	31	306
Fusarium oxysporum F3	3	25	3	38	1219
Penicillium purpurogenum	5	5.7	15.8	33	966
Bacillus circulans	3	29.5	38.5	30	2039.5
Based on Sa-Pereira et al., 2003					

indicating that the enzyme was purified to apparent homogeneity, while zymography showed a corresponding sharp activity band corresponding to a 42-kDa protein estimated from the migration of molecular weight standards after electrophoresis (Table 6.2).

REFERENCES

Abirami, V., Meenakshi, S.A., Kanthymathy, K., Bharathidasan, R., Mahalingam, R., Panneerselvam, A., 2011. Partial purification and characterization of an extracellular xylanase from *Penicillium janthinellum* and *Neurospora crassa*. J. Nat. Prod. Plant Res. 1 (4), 117−125.

Bajpai, P., 2009. Xylanases. In: Schaechter, M., Lederberg, J. (Eds.), Encyclopedia of Microbiology, Third ed. Academic Press, San Diego, pp. 600−612.

Breccia, J.D., Kaul, R.H., Castro, G.R., Sineriz, F., 1999. Isolation of β-xylanase from whole broth of *Bacillus amyloliquefaciens* by adsorption on a matrix in fluidized bed with low degree of expansion. Bioseparation 8, 273−279.

Bronnenmeier, K., Adelsberger, H., Lottspeich, F., Staudenbauer, W.L., 1996. Affinity purification of cellulose-binding enzymes of *Clostridium stercorarium*. Bioseparation 6, 41−45.

Carmona, E.C., Fialho, M.B., Buchgnani, E.B., Coelho, G.D., Brocheto-Braga, M.R., Jorge, J. A., 2005. Production, purification and characterization of a minor form of xylanase from *Aspergillus versicolor*. Process. Biochem. 40, 359−364.

Chandra, K.R., Chandra, T.S., 1996. Purification and characterization of xylanase from alkali-tolerant *Aspergillus fischeri* Fxn1. FEMS Microbiol. Lett. 145, 457−461.

Chapla, D., Patel, H., Singh, A., Madamwar, D., Shah, A., 2011. Production, purification and properties of a cellulase-free thermostable endoxylanase from newly isolated *Paenibacillus* sp. ASCD2. Ann. Microbiol. Available from: http://dx.doi.org/10.1007/s13213-011-0323-5.

Dhillon, A., Gupta, J.K., Khanna, S., 2000. Enhanced production, purification and characterization of a novel cellulase-poor thermostable, alkali tolerant xylanase from *Bacillus circulans* AB 16. Process. Biochem. 35, 849−856.

Fengxia, L., Mei, L., Zhaoxin, L., Xiaomei, B., Haizhen, Z., Yi, W., 2008. Purification and characterization of xylanase from *Aspergillus ficuum* AF-98. Bioresour. Technol. 99, 5938−5941.

Fontes, CMGA, Gilbert, H.J., Hazlewood, G.P., Clarke, J.H., Prates, J.A.M., Mckie, V.A., et al., 2000. A novel cell vibrio mixtus family 10 xylanase that is both intracellular and expressed under noninducing conditions. Microbiology 146, 1959−1967.

Gupta, M.N., Guoqiang, D., Kaul, R., Mattiasson, B., 1994. Purification of xylanase from Trichoderma viride by precipitation with an anionic polymer Eudragit S 100. Biotechnol. Tech. 8, 117–122.

Irfan, M., Syed, Q., 2012. Partial purification and characterization of xylanase from Trichoderma viride produced under SSF. Int. J. Appl. Res. Nat. Prod. 5 (1), 7–11.

Jorgensen, H., Eriksson, T., Borjesson, J., Tjerneld, F., Olsson, L., 2003. Purification and characterization of five cellulase and one xylanase from Penicillium brasilianum IBT 20888. Enzyme Microb. Technol. 32, 851–861.

Kamble, R.D., Jadhav, A.R., 2012. Isolation, purification, and characterization of xylanase produced by a new species of Bacillus in solid state fermentation. Int. J. Microbiol., 1–8.

Kiddinamoorthy, J., Anceno, A.J., Haki, G.D., Rakshit, S.K., 2008. Production, purification and characterization of Bacillus sp. GRE7 xylanase and its application in eucalyptus kraft pulp biobleaching. World J. Microbiol. Biotechnol. 24, 605–612.

Kumar, S.S., Panday, D.D., Naik, G.R., 2011. Purification and molecular characterization of low molecular weight cellulase-free xylanase from thermoalkalophilic bacillus spp. World J. Sci. Technol. 1 (2), 9–16.

Lappalainen, A., Sikka-Aho, M., Kalkkinen, N., Fagerstorm, R., Tenkanen, M., 2000. Endoxylanases II from Trichoderma reesei has several isoforms with different isoelectric points. Biotechnol. App. Biochem. 59, 61–68.

Lee, Y.E., Lowe, S.E., Zeikusi, J.G., 1993. Regulation and characterization of xylanolytic enzymes of Thermoanaerobacterium saccharolyticum B6A-RI. Appl. Environ. Microbiol. 59, 763–771.

Li, X.T., Jiang, Z.Q., Li, L.T., Yang, S.Q., Feng, W.Y., Fan, J.Y., et al., 2005. Characterization of a cellulase-free, neutral xylanase from Thermomyces lanuginosus CBS 288.54 and its biobleaching effect on wheat straw pulp. Bioresour. Technol. 96, 1370–1379.

Mamo, G., Hatti-Kaul, R., Mattiasson, B., 2006. A thermostable alkaline active endo-β-1,4-xylanase from Bacillus halodurans S7: purification and characterization. Enzyme Microb. Technol. 39, 1492–1498.

Matte, A., Forsberg, C.W., 1992. Purification, characterization, and mode of action of endoxylanases 1 and 2 from Fibrobacter succinogenes S85. App. Environ. Microbiol. 59, 157–168.

Milagres, A.M.F., Magalhaes, P.O., Ferraz, A., 2005. Purification and properties of a xylanase from Ceriporiopsis subvermispora cultivated on Pinus taeda. FEMS Microbiol. Lett. 253, 267–272.

Nakamura, S., Wakabayashi, K., Nakai, R., Aono, R., Horikoshi, K., 1993. Purification and some properties of an alkaline xylanase from alkaliphilic Bacillus sp., strain 41M-1. Appl. Environ. Microbiol. 59, 2311–2316.

Ninawe, S., Kapoor, M., Kuhad, R.C., 2008. Purification and characterization of extracellular xylanase from Streptomyces cyaneus SN32. Bioresour. Technol. 99, 1252–1258.

Palmer, T., 2001. Enzymes: Biochemistry. Biotechnology and Chemical Chemistry. Horwood Publication, Chicester, UK, pp. 191–222.

Rao, M., Khadilkar, S., Bandivadekar, K.R., Deshpande, V., 1996. Structural environment of an essential cysteine residue of xylanase from Chainia sp. Biochem. J. 316, 771–775.

Rodrigues, E.M.G., Tambourgi, E.B., 2001. Continuous extraction of xylanase from Penicillium janthinellum with reversed micelles using experimental design mathematical model. Biotechnol. Lett. 56, 765–766.

Roy, N., Habib, R.M., 2009. Isolation and characterization of xylanase producing strain of Bacillus cereus from soil. Iranian J. Microbiol. 1 (2), 49–53.

Sandrim, V.C., Rizzatti, A.C.S., Terenzi, H.F., Jorge, J.A., Milagres, A.M.F., Polizeli, MLTM, 2004. Purification and biochemical characterization of two xylanases produced by Aspergillus caespitosus and their potential for kraft pulp bleaching. Process Biochem. 40, 1823–1828.

Santos, E.S.D., Guirardello, R., Franco, T.T., 2002. Preparative chromatography of xylanase using expanded bed adsorption. J. Chromat. A 944, 217–224.

Sa-Pereira, P., Paveia, H., Costa-Ferreira, M., Aires-Barros, M.R., 2003. A new look at xylanases: an overview of purification strategies. Mol. Biotechnol. 24, 257–281.

Shrinivas, D., Savitha, G., Raviranjan, K., Naik, G.R., 2010. A highly thermostable alkaline cellulase-free xylanase from thermoalkalophilic Bacillus sp. JB 99 suitable for paper and pulp industry: purification and characterization. Appl. Biochem. Biotechnol. 162, 2049–2057.

Subramaniyan, S., Prema, P., 2002. Biotechnology of microbial xylanases: enzymology, molecular biology and application. Crit. Rev. Biotechnol. 22, 33–46.

Teplitsky, A., Shulami, S., Moryles, S., Shoham, G., 2000. Crystallization and preliminary X-ray analysis of an intracellular xylanase from *Bacillus stearothermophilus* T-6. Acta Crystallogr. D Biol. Crytallogr. 56, 181–186.

Wong, K.K.L., Tan, L.U.L., Saddler, J.N., 1988. Multiplicity of β-1-4-xylanase in microorganisms: functions and applications. Microbiol. Rev. 52, 305–317.

Wong, K.K.Y., Saddler, J.N., 1992. Trichoderma xylanases, their properties and purification. Crit. Rev. Biotechnol. 12, 413–435.

Yasinok, A.E., Biran, S., Kocabas, A., Bakir, U., 2010. Xylanase from a soil isolate, *Bacillus pumilus*: gene isolation, enzyme production, purification, characterization and one-step separation by aqueous-two-phase system. World J. Microbiol. Biotechnol. 26, 1641–1652.

Zhiwei, L., Yang, J., Yuan, H., 2008. Production, purification and characterization of an alkaliphilic endo-β-1,4-xylanase from a microbial community EMSD5. Enzyme Microb. Technol. 43, 343–348.

CHAPTER 7

Immobilization of Xylanases

Enzymes display great specificity and are not permanently modified by their participation in reactions. Since they are not changed during the reactions, it is cost-effective to use them more than once. Generally, the free forms of enzymes pose difficulties during their application with respect to instability of enzyme structure and sensitivity under harsh conditions, nonrecovery of the active form of the enzyme from the reaction mixtures, and contamination problems (Price and Stevens, 1999). Thus, the most obvious reason for immobilization is the need to reuse enzymes if they are expensive in order to make their use in industrial processes economical. Moreover, it is also believed that the stability of enzymes increases upon immobilization (Palmer, 2000).

The major advantages of immobilizing enzymes are: the multiple or repetitive uses of a single batch of enzymes, the ability to stop the reaction rapidly by removing the enzyme from the reaction system, the stabilization of the enzyme by bounding (usually), and the absence of contamination of the product by the enzyme. Immobilizing enzymes also possesses analytical advantages, such as: a long half-life, predictable decay rates, the elimination of a reagent preparation, and so forth. Immobilization of enzymes also has certain limitations, such as: the high cost of carriers for immobilization, problems with cofactors and regeneration, problems with multienzyme systems, changes in the properties of enzymes, and activity loss during immobilization, among others. Immobilization of an enzyme results in altering its catalytic activity or thermal stability (Brena and Viera, 2006). Immobilization of enzymes can also provide increased resistance to changes in fluctuating conditions, such as pH or temperature. It allows enzymes to be held in place throughout the reaction. Immobilized enzymes can easily be separated from products after reactions, so they can be used again and again. Therefore, immobilization of enzymes increases catalytic effects and is widely used in industry for enzyme-catalyzed reactions (Pal and Khanum, 2011). The benefits of the immobilization of enzymes are summarized in Table 7.1.

Table 7.1 Benefits of Immobilized Enzyme
• Multiple or repetitive use of a single batch of enzymes
• The ability to stop the reaction rapidly by removing the enzyme from the reaction solution (or vice versa)
• Enzymes are usually stabilized by binding
• Product is not contaminated with the enzyme (especially useful in the food and pharmaceutical industries)
• Analytical purposes—long half-life, predictable decay rates, elimination of reagent preparation, etc.

Alginic acid plays an important role in the immobilization of microbial enzymes in many industrial processes (Arya and Srivastava, 2006; Goel et al., 2006; Pal and Khanum, 2012).

There are several methods of immobilizing the enzyme depending upon the end use required. When immobilizing an enzyme to a surface, it is most important to choose a method of attachment that will prevent loss of enzyme activity by not changing the chemical nature or reactive groups in the binding site of the enzyme. The rate and yield of immobilization depends on parameters such as: type of carrier, method of immobilization, concentration, pH, temperature, and reaction time. In immobilization studies, either the whole cell immobilization of an organism to a solid support is performed or sometimes the enzyme, itself, is immobilized on a reversible soluble–insoluble polymer.

Adsorption is the best-established technique for the noncovalent immobilization of enzymes. It is based on the physical adsorption or ionic binding or both of the enzyme to the surface of the support. This is a low cost, simple, and very effective procedure. Some enzymes can be immobilized by adsorption on DEAE-Sephadex, alumina, or charcoal, to name a few. The entrapment of an enzyme can be done on various polymers. The matrices commonly used have lattice dimensions that allow easy molecular migration. Thus, to maximize enzyme entrapment and to minimize leakage, a high degree of cross-linking is required by the polymer. Some examples of such polymers are PAGE, fibers, sodium alginate, and agar. Encapsulation of enzymes can be done in liposomes or hollow fiber membranes, to name only a few examples. The technique is simple and chemical reactions are not required. The membranes confine macromolecules, but allow small molecules to pass through. The main membranes used are polysulfones, cellulose acetate, and acrylic copolymers. The major problem related to such systems is membrane fouling.

Many research groups have immobilized xylanase in different manners to increase efficiency. *Thermotoga maritima* xylanase B was immobilized using reversibly irreversible polymer Eupergit C, and it showed great thermal stability, maintaining 92.5% of its activity even after 10 cycles at 90°C with a half-life of 24 days (Tan et al., 2008). Ai et al. (2005) have reported that immobilized xylanase on Eudragit S-100 from *Streptomyces olivaceoviridis* produced 15.5 mg/ml XOS from pretreated corncobs at 24 h of reaction time. Kapoor and Kuhad (2007) immobilized xylanase from *Bacillus pumilus* by: entrapment in polyacrylamide or calcium alginate; adsorption on silica and alumina or chitin; and ionic binding with DEAE sepharose or Amberlite IR-120. Guerfali et al. (2009) immobilized *Talaromyces thermophilus* β-xylosidase and observed its improvement in its catalytic properties for the production of xylose and xylooligosaccharides. They used different matrices for immobilization, such as: DEAE-cellulose and DEAE-sephadex for ionic binding; polyacrylamine for entrapment; and covalent coupling using chitosan, chitin, and gelatin.

Madakbas et al. (2013) studied immobilization of xylanase by using activated polyaniline. Polyaniline was selected because of its low cost and promising electrical and optical properties (Amerithesh et al., 2008). They used powdered polyaniline to immobilize the xylanase via glutaraldehyde by covalent binding. The optimum pH of the immobilized enzyme was shifted toward the acidic region by 1.0 pH unit. The storage stability and reusability of the immobilized xylanase were also improved and found to be greater than that of the free xylanase enzyme.

Gwande and Kamat (1998a) immobilized *Aspergillus sp.* strain 5 and *Aspergillus sp.* strain 44 on 400-mesh nylon bolting cloth in a shake flask culture. They reported a 1.68-fold higher xylanase yield in immobilized *Aspergillus sp.* strain 5 than that of freely suspended cells. The xylanase from the same organism was noncovalently immobilized on Eudragit S-100 for saccharification, which enabled its recovery and reuse for a longer period (Gwande and Kamat, 1998b). Tokuda et al. (1997) also reported that maximum xylanase yield can be obtained by immobilizing *Aspergillus niger* on silk, rayon, and polyester fibers, which have several advantages over free enzymes. In one study (Beg et al., 2000), the effectiveness of polyurethane foam (PUF) and three nonwoven fabrics, namely cotton, silk, and polyester, as support materials for *Streptomyces sp.* QG-11-3 mycelia immobilization was investigated. The xylanase yields were enhanced by 2.5-fold, 1.91-fold, 1.54-fold, and 1.47-fold using PUF, polyester, silk, and cotton, respectively, compared with the xylanase yield in liquid-batch fermentation. These results also indicated that the type of fiber material has a significant role in providing a favorable environment for enzyme production, thus having an influence on the xylan hydrolysis activity of the immobilized mycelia of *Streptomyces sp.* QG-11-3. Mycelia grew inside the pores of fabric material and PUF particles. Unlike other techniques involving active immobilization, the use of PUF particles does not require the growth of cells prior to immobilization. The inert particles are simply placed in the fermenter before sterilization and the fermenter is inoculated in the normal way. Mycelia/cells become immobilized within the PUF pores as a natural consequence of growth, during an initial growth period. This technique has also been applied successfully to a wide variety of microbial cell systems for immobilization.

Amani et al. (2007) immobilized *Bacillus pumilus* cells by using agar and alginate gel materials. Immobilization of cells in 2% alginate showed a 2.6-fold increase in xylanase specific activity compared to free cells. Hydrogels comprising polyanion xylan and polycation chitosan were used as carriers for immobilization of the endoxylanase from *T. viride* alone or together with the XIX type protease from *A. sojae*. Due to the induced electrostatic interaction between the enzyme and gel, the immobilized endoxylanase was active at 98°C, while the free enzyme was active only up to 45°C. In the case of the co-immobilized endoxylanase and protease, due to the synergistic interactions between the two enzymes, the protease activity was also drastically increased (Dumitriu et al., 1997).

Efficient immobilization of endoxylanase over siliceous supporting materials by three steps was described. Siliceous material was incubated with the cross-linking agent polyaldehyde followed by mixing the enzyme with the cross-linked material to bind the enzyme on it. The immobilized enzyme was separated from the free enzyme by filtration, achieving 40% of endoxylanase transferred to the support material (Le Fevre and Saville, 1999).

β-Xylosidase from *A. niger* was immobilized by adsorption on a polyamide membrane with 30% of the enzyme being immobilized. The immobilized enzyme kept only 6.8% of the maximal activity of the free enzyme, but showed a better thermal stability than the free enzyme.

The immobilized enzyme was still active after 20 cycles of usage for xylan hydrolysis, releasing xylose at $1.7 \, g \, l^{-1}$ enzyme (Delcheva et al., 2008).

REFERENCES

Ai, Z., Jiang, Z., Li, L., Deng, W., Kusakabe, I., Li, H., 2005. Immobilization of *Streptomyces olivaceoviridis* E-86 xylanase on Eudragit S-100 for xylo-oligosaccharide production. Process Biochem. 40, 2707–2714.

Amani, M.D., Ahwany, E., Amany, S.Y., 2007. Xylanase production by *Bacillus pumilus*: optimization by statistical and immobilization methods. Res. J. Agric. Biol. Sci. 3, 727–732.

Amerithesh, M., Aravind, S., Jayalekshmi, S., Jayasree, R.S., 2008. Polyaniline doped with orthophosphoric acid—a material with prospects for optoelectronic applications. J. Alloys Compd. 458, 532–535.

Arya, S.K., Srivastava, S.K., 2006. Kinetics of immobilized cyclodextrin gluconotransferase produced by *Bacillus macerans* ATCC 8244. Enzyme. Microb. Technol. 39, 507–510.

Beg, Q.K., Bhushan, B., Kapoor, M., Hoondal, G.S., 2000. Enhanced production of a thermostable xylanase from Streptomyces sp. QG-11-3 and its application in biobleaching of eucalyptus kraft pulp. Enzyme. Microb. Technol. 27, 459–466.

Brena, B.M., Viera, F.B., 2006. Immobilization of enzymes—a literature survey. Methods Biotechnol. 22, 15–30.

Delcheva, G., Dobrev, G., Pishtiyski, I., 2008. Performance of *Aspergillus niger* B 03 β-xylosidase immobilized on polyamide membrane support. J. Mol. Catal. B Enzym. 54, 109–115.

Dumitriu, S., Chornet, E., Vidal, P., 1997. Polyionic insoluble hydrogels comprising xanthan and chitosan. US Patent 5620706.

Goel, A., Sharma, R.K., Tandon, H.K.L., 2006. A comparison of different polymeric gels for entrapment of cells of *Streptococcus thermophilus* containing β-galactosidase. J. Food Sci. Technol. 43 (5), 526–531.

Guerfali, M., Maalej, I., Gargouri, A., Belghith, H., 2009. Catalytic properties of the immobilized Talaromyces thermophilus β-xylosidase and its use for xylose and xylooligosaccharides production. J. Mol. Catal B-Enzyme. 57, 242–249.

Gwande, P.V., Kamat, M.Y., 1998a. Immobilization of *Aspergillus* sp. on nylon bolting cloth for production of xylanase. J. Ferment Bioeng. 86, 243–246.

Gwande, P.V., Kamat, M.Y., 1998b. Production, characterization and application of *Aspergillus* sp. xylanase immobilized on Eudragit S-100. J. Biotechnol. 66, 165–175.

Kapoor, M., Kuhad, R.C., 2007. Immobilization of xylanase from *Bacillus pumilus* strain MK001 and its application in production of xylo-oligosaccharides. Appl. Biochem. Biotechnol. 142, 125–138.

Le Fevre, G.N., Saville, B.A., 1999. Enzyme immobilization on a siliceous support with a polyaldehyde cross-linking agent. US Patent 5998183.

Madakbas, S., Danis, O., Demir, S., Kahraman, M.V., 2013. Xylanase immobilization on functionalized polyaniline support by covalent attachment. Starch 65, 146–150.

Pal, A., Khanum, F., 2011. Covalent immobilization of xylanase on glutaraldehyde activated alginate beads using response surface methodology: characterization of immobilized enzyme. Process Biochem. 46, 1315–1322.

Pal, A., Khanum, F., 2012. Covalent immobilization of xylanase on the surface of alginate glutaraldehyde beads decreases the 'catalytic efficiency' but provides 'low temperature stabilization' effect. J. Biochem. Technol. 3 (4), 409–413.

Palmer, T., 2000. Kinetics of single-substrate enzyme-catalyzed reactions. In: Enzymes: Biochemistry, Biochemistry and Clinical Chemistry, pp. 107–113.

Price, N.C., Stevens, L., 1999. Enzyme technology. In: Fundamentals of Enzymology: The Cell and Molecular Biology of Catalytic Proteins, third ed. Oxford University Press, UK, pp. 430–449.

Tan, S.S., Li, D.Y., Jiang, Z.Q., Zhu, Y.P., Shi, B., Li, L.T., 2008. Production of xylobiose from the autohydrolysis explosion liquor of corncob using *Thermotoga maritima* xylanase B (Xyn B) immobilized on nickel-chelated Eupergit C. Bioresour. Technol. 99, 200–204.

Tokuda, H., Urata, T., Nakanishi, K., 1997. Hydrolysis of xylan by *Aspergillus niger* immobilized on nonwoven fabrics. Biosci. Biotechnol. Biochem. 61, 583–589.

CHAPTER 8

Industrial Applications of Xylanases

Xylanases have stimulated great interest due to their potential application in several industries (Bajpai, 1997, 1999, 2009, 2012; Beg et al., 2001; Bajpai et al., 2004; Polizeli et al., 2005; Butt et al., 2008; Dhiman et al., 2008a; Harris and Ramalingam, 2010; Juturu and Wu, 2012; Sharma and Kumar, 2013). In recent years, the biotechnological use of xylans and xylanases has grown remarkably. Xylanase began to be used in the 1980s, initially in the preparation of animal feed and later in the food, textile, paper, and chemical industries. Polizeli et al. (2005) reported that xylanase and cellulase, together with pectinases, account for 20% of the world enzyme market.

8.1 PULP AND PAPER

8.1.1 Prebleaching of Pulp

Use of enzymes in pulp bleaching has attracted attention in recent years (Viikari et al., 1986, 1990, 1991, 1993, 1994, 2002, 2009; Bajpai et al., 1993, 1994, 1999; Bajpai, 1997, 1999, 2004, 2009, 2012, 2013a; Bajpai and Bajpai, 1997, 2001a; Ojanpera, 2004; Paice and Zhang, 2005; Manji, 2006). Enzymes of the hemicellulolytic type, particularly xylanases, are used commercially for pulp bleaching. Xylanase bleaching of chemical pulp is the most widely used and best established biotechnical application in pulp bleaching. This technology has become one of the solutions considered by the pulp and paper industry to give an innovative, environmentally and economically acceptable answer to the pressures exerted on chlorine bleaching by regulatory authorities in Western countries and by more demanding, environmentally minded consumers (Viikari et al., 2002, 2009; Bajpai, 2004; Reid et al., 2010). By use of xylanase enzymes, a variety of bleaching benefits can be obtained. These include:

- Reduction of AOX discharges, primarily by decreasing chlorine gas usage;
- Debottlenecking mills limited by chlorine dioxide generator capacity;

- Elimination of chlorine gas usage for mills at high chlorine dioxide substitution levels;
- Increasing the brightness ceiling, particularly for mills contemplating ECF and TCF bleaching sequences;
- Decreasing cost of bleaching chemicals, particularly for mills using large amounts of peroxide or chlorine dioxide.

These benefits are achieved over the long term when the enzymes are selected and applied properly in the mill (Viikari et al., 1994).

The main enzyme needed to enhance the delignification of kraft pulp is reported to be endo-β-xylanase, but enrichment of xylanase with other hemicellulolytic enzymes has been shown to improve the effect of enzymatic treatment. The use of xylanase enzymes to enhance the bleaching of the pulp was first reported in 1986 by Viikari et al. The Finnish forest companies were the first in the world to start mill-scale trials in 1988. In North America, the first mill trials of xylanase were carried out at Port Alberni in 1991 and ongoing usage in some mills started in 1992. Presently, a significant number of European, North American, South American, and Japanese mills are bleaching full time with enzymes. Xylanase use is more common in Canada than the U.S. because of more stringent AOX levels. Various paper products, including magazine paper and tissue papers manufactured from xylanase-treated pulp, have been successfully introduced to the market. In the U.S., promulgation of the Cluster rule pressured mills to complete conversion to elemental chlorine-free bleaching in an economical way. In addition, weak markets for many pulp and paper products have forced mills to lower their bleaching costs. Fortunately, the use of xylanase enzymes can address both needs. Most full-time applications focus on cost reductions using O_2-ECF bleaching. Recent developments involve use of enzymes to eliminate the first chlorine dioxide stage and thereby help reduce water usage. In 2005, about 10% of all kraft pulp was manufactured with xylanase prebleaching. In North America, Iogen Corp—based in Ottawa—has established a market leadership position. Globally, other suppliers such as AB Enzymes and, more recently, Diversa are also selling to this market. In Japan, Oji Paper is unique in manufacturing xylanase onsite at its Yonago mill (Bajpai, 2012).

Xylanase applications are often referred to as prebleaching or bleach boosting because the nature of the effect is to enhance the effect of bleaching chemicals rather than to remove lignin directly. The enzyme does not attack the lignin-based chromophores, but rather the xylan network by which the residual lignin particles are surrounded and trapped. A limited hydrolysis of the xylan network is often sufficient to facilitate the subsequent chemical attack on lignin with various bleaching chemicals, without sacrificing yield. The results from laboratory studies and mill trials show about a 35–41% reduction in active chlorine at the chlorination stage for hardwoods and 10–20% for softwoods, whereas the savings in total active chlorine were found to be 20–25% for hardwoods and 10–15% for softwoods (Senior and Hamilton, 1991, 1992a, b, c; Senior et al., 1992, 1999, 2000; Skerker et al., 1992; Tolan, 1992, 2001; Tolan and Canovas, 1992; Yang and Eriksson, 1992; Allison et al., 1993a, b; Atkinson et al. 1993; Bajpai et al., 1993, 1994, 1999; Werthemann, 1993; Tolan et al., 1996; Tolan and Guenette, 1997; Valchev et al., 1998, 2000; Awakaumova et al., 1999; Thibault et al., 1999; Bim and Franco, 2000; Zhan et al., 2000; Saleem et al., 2009; Ko et al., 2011). In the elementary chlorine-free bleaching sequences, the use of xylanase increases the productivity of the bleaching plant when the production capacity of ClO_2 is a limiting factor. This is often the case when the use of chlorine gas has been abandoned. In totally chlorine-free (TCF) bleaching sequences, the addition of xylanase increases the final brightness value, which is a key parameter in the marketing of chlorine-free pulps. In addition, the savings in TCF bleaching are important with respect both to costs and to the strength properties of the pulp. Xylanase pretreatment leads to reductions in effluent AOX and dioxin concentrations due to the reduced amount of chlorine required to achieve a given brightness. The level of AOX in effluent is significantly lower for xylanase-pretreated pulps as compared to conventionally bleached control pulps. The xylanase-treated pulps show unchanged or improved strength properties. Also, these pulps are easier to refine than the control pulps. The viscosity of the pulp is improved as a result of xylanase treatment. However, the viscosity of the pulp is adversely affected when cellulase activity is present. Therefore, the presence of cellulase activity in the enzyme preparation is not desirable.

The actual enzymatic mechanism of xylanase prebleaching is not yet well understood. One hypothesis suggests that precipitated xylan blocks or occludes extraction and that xylanase increases the accessibility (Kantelinen et al., 1993). This model is based on reports that xylan reprecipitates on the fiber surfaces (Yllner et al., 1957). It has been reported that no extensive relocation of xylan to the outer surface occurs during pulping, so the occlusion model might not be a sound premise (Suurnakki et al., 1997). Another possible explanation for xylanase action in bleaching is that the disruption of a xylan chain by xylanase interrupts lignin−carbohydrate bonds, improves the accessibility of the bleaching chemicals to the pulps, and facilitates easier removal of solubilized lignin in bleaching (Paice et al., 1992). Skjold-Jorgensen et al. (1992) found that xylanase treatment decreased the demand for active chlorine for a batch kraft pulp by 15%, but decreased active chlorine of pulp from a continuous process by only 6 to 7%. They also showed that DMSO extraction of residual xylan does not lead to an increase in bleachability, but that xylanase treatment does. This shows that DMSO-extractable xylan is not involved in bleach boosting. Paice et al. (1992) have shown that the prebleaching effect on black spruce pulp is associated with a drop in the degree of polymerization, even though the xylan content decreases slightly. Prebleaching thus appears to be associated with xylan depolymerization, even though not necessarily with solubilization of the xylan-derived hemicellulose components. Senior and Hamilton (1993) have shown that xylanase treatment and extraction change the reactivity of the pulp by enabling a higher chlorine dioxide substitution to achieve a target brightness and that they raise the brightness ceiling of fully bleached pulps.

Canadian researchers showed that xylanase treatment and extraction change the reactivity of the pulp by enabling a higher chlorine dioxide substitution to achieve a target brightness and that they raise the brightness ceiling of fully bleached pulps. Xylanase enzyme was used in a Canadian pulp mill to reduce the consumption of chlorine and chlorine dioxide in the bleaching processes (Manji, 2006). The total equivalent chlorine decreased by 8 kg/air dried (a.d.) metric ton of pulp during enzyme treatment. Consequently, the substitution in the front end of the bleach plant increased from 28.5% to 36.4%. In addition, the active chlorine multiple decreased from 0.23 to 0.21 without a loss in pulp production rate and with an insignificant change in the physical properties of pulp. During the enzyme treatment period, the AOX being discharged into the receiving waters decreased from 2.4 kg/a.d. metric ton to 2.2 kg/a.d. metric ton. The pulp quality results showed no significant difference in the strength factor during the enzyme treatment period.

Shatalov and Pereira (2009) observed an increase in brightness and a reduction in hexenuronic acid (Hex-A) of E. globules kraft pulp treated with commercial xylanase enzymes: Ecopulp and Pulpzyme in single-, two-, and three-stage peroxide bleaching (Tables 8.1 and 8.2). The effect of various cellulase-free commercial xylanases in prebleaching of bamboo kraft pulp was reported by Bajpai and Bajpai (1996) in C_DEHD sequence (Table 8.3). The enzyme-treated pulps showed higher brightness in comparison to controls.

Shatalov and Pereira (2009) used chemical pulps to see the impact of hexenuronic acids on xylanase prebleaching and they found that xylanase assisted in direct pulp brightening and suggested that it is due to Hex-A removal with solubilized xylooligosaccharide fractions. Strong positive correlation was established between the xylanase bleach boosting effect and the bleaching profile of Hex-A. Wong et al. (2001) studied the effect of alkali and oxygen extractions of kraft pulp on xylanase prebleaching and they observed improvement in final brightness by up to 1.4−2.1 unit with reduction in Hex-A and kappa number of xylanase-pretreated pulps than the corresponding control pulps.

The most conventional method is to add xylanase to the brownstock pulp prior to the high density (HD) tower (Tolan and Canovas, 1992; Tolan et al., 1996; Tolan and Guenette, 1997; Tolan, 2001). The enzyme reaction takes place in the tower and the treated pulp then

Table 8.1 Effect of Xylanase Aided Three-stage Hydrogen Peroxide Bleaching on Brightness of E. globulus Kraft Pulp*

Bleaching Stage		Brightness (%ISO)		Hexenuronic Acid (μmol/g dry pulp)	
		Ecopulp	Pulpzyme	Ecopulp	Pulpzyme
X	Control	42.4	42.0	50.45	49.16
	Enzyme	43.9	43.2	43.13	43.69
XQP	Control	75.8	74.8	44.13	44.04
	Enzyme	77.9	76.9	34.69	35.92
XQPP	Control	81.2	80.1	39.81	40.19
	Enzyme	82.9	81.5	32.19	33.77
XQPPP	Control	85.0	84.5	36.60	37.14
	Enzyme	86.4	85.7	30.63	31.88

*X − Enzymatic pretreatment; Q − Chelating; P − Peroxide bleaching stage
Based on data from Shatalov and Pereira, 2009

Table 8.2 Increase in the Brightness of Pulps by various Cellulase-free Commercial Xylanases Using C$_D$EHD Bleaching Sequence*

Particular	Reaction pH	Reaction Temp. (°C)	Enzyme Dose (IU/g)	Brightness (%ISO)	Increase in Brightness, Points
Control	—	—	—	87.1	—
Bleachzyme B	7–7.5	40–50	10.0	87.9	0.8
Bleachzyme F	6–6.5	45–50	6.0	88.6	1.5
VAI xylanase	5–7	50–60	3.5	88.4	1.3
Ecopulp-X200	5–6	50–55	10.0	88.5	1.4
Cartazyme HS-10	4–5	40–55	13.0	88.6	1.5
Pulpzyme HC	8–9	60–70	14.0	88.0	0.9
Pulpzyme HB	7–8	45–55	14.0	88.2	1.1
Irgazyme-10	5.5–6	40–50	12.0	88.5	1.4
Irgazyme 40S	7–8	50–60	7.5	88.6	1.5

*C$_D$ – Chlorination stage; E – Extraction; H – Hypo; D – chlorine dioxide
Based on data from Bajpai and Bajpai 1996

passes into the bleach plant. Various ways to add enzymes have been used. These include:

1. Spraying on the decker pulp mat,
2. Adding to either the decker repulper or discharge chute,
3. Adding into the stock of medium consistency pulp leading to the HD tower,
4. Adding directly into the HD tower.

Xylanase has also been added later in the bleaching sequence, rather than to the brownstock pulp. The latest generation of alkali tolerant enzymes require little, if any, addition of acid to adjust the pH. Earlier generation of enzymes had pH optima ranging from 5 to 6.5 and required acid addition to brownstock pulp. Instances of corrosion problems were seen when acid was incorrectly applied. New xylanases have higher pH optima and function optimally without pH adjustment.

The acid of preference by far has been sulfuric acid. However, with the development of alkaline xylanases, noncorrosive carbon dioxide is an excellent choice and also improves washer performance. The addition of acid prior to the D$_0$ stage in ECF bleaching has also been shown to improve the performance of the D$_0$ stage. This is because the higher acidity in the stage prevents the decomposition of chlorine dioxide to chlorate and because the chemistry of delignification with chlorine dioxide favours an acidic environment. Acid added to the low consistency pulp prior to the washer vat provides the benefits of reducing pitch deposits; however, acid charges here tend to be much higher due to the large volume that must be treated. Acid can also be added on the washer shower, bars shower, or in the repulper discharge section. Experience has shown that the prevention of corrosion must be a priority.

Detailed laboratory work is generally needed to optimize and adapt the enzymatic treatment to individual existing mill conditions. Interestingly, however, xylanase bleaching has been scaled up directly from laboratory scale to the large industrial scale (1000 TP/d) without intermediate pilot stages. It has also been observed that even higher brightness values can be reached on the full scale than those attainable in the laboratory, which is due to the more efficient mixing systems and higher pulp consistencies. No expensive capital investments have generally been necessary for full-scale runs. The most significant requirement is the addition of pH adjustment facilities. Xylanase pretreatment has been shown to be easily applicable with existing industrial equipment, which is a considerable advantage of this technology.

Bleaching with xylanase requires proper control of pH and temperature as well as retention time (Tolan and Guenette, 1997). The optimum pH and temperature for enzyme treatment varies among enzymes. Generally, xylanases derived from strains of bacterial origin are most effective between pH 6 and 9, while those derived from strains of fungal origin should be used within the pH range of 4 to 6. The optimum temperature ranges from 35°C to 60°C for different enzymes. To obtain the best results from enzyme use, enzyme dosage must be optimized in each single case. In addition, the pulp consistency must be optimized to obtain effective dispersion of the enzyme and to improve the efficiency of the enzyme treatment. Screw conveyers and static mixers are examples of efficient mixing systems. Most of the bleaching effect is obtained after only one hour of treatment. Usually, the reaction time is set to 2–3 hours. Long reaction time must be avoided if cellulases are present. Commercial xylanase enzyme preparations consist mainly of endoxylanases. Most of the enzymes are active at acidic or neutral pH, although some of them function under alkaline conditions. Xylanases are sold as concentrated liquids and the amount required per metric ton of pulp is very low, less than a liter. The cost of enzyme per ton of pulp varies and depends on the dosage required and the supplier. The approximate cost of an enzyme treatment is around $1.2 to 2.0/TP. Due to the low enzyme price and the low capital costs of the enzyme stage, the potential economic benefits of enzyme bleaching are significant.

Mill operations also affect the performance of the xylanase enzyme. The effect of raw material, the pulping process, brown stock washing, and the bleaching sequence should be assessed by laboratory testing prior to mill usage of enzymes.

Among raw materials, the important distinction is between hardwoods and softwoods. The percentage of the bleaching chemicals saved by xylanase treatment is greater on hardwoods than on softwoods. At good treatment conditions, the decrease in chlorine chemicals is about 20% on hardwoods and 15% on softwoods. The digestor operation affects the xylan content of the pulp significantly. For example, sulfite pulping destroys most of the xylan and thus sulfite pulp is not suitable for enhanced bleaching by enzyme treatment. In conventional kraft pulping, the xylan content depends strongly on the effective alkalinity. The lower the alkalinity, the higher the xylan content and the benefits of using xylanase enzymes. At high alkalinity (19−22%) much of the xylan is solubilized, which decreases the benefit of a xylanase treatment. At low alkalinity (less than about 18%), the xylan structure is more stable, and the bleaching enhancement by xylanase is greater by up to 2- to 3-fold over the high alkalinity pulp. This is often the case for pulp cooked to a higher kappa number. Kraft pulping at severe conditions, such as conventional cooking of softwood to kappa number less than 23, also destroys much of the hemicellulose that is accessible to the enzyme. On the other hand, MCC or oxygen-delignified pulps with low unbleached kappa number respond well to enzyme treatment. Much smaller enzyme benefit has been reported for batch-cooked pulp at kappa number 21 than for MCC and oxygen-delignified pulp at the same kappa number. The MCC and oxygen-delignified pulps have hemicellulose structures that are similar to that for conventional, high kappa number pulps. Enzyme benefits have been achieved in mills with conventional, MCC, and O_2 delignification systems. The brownstock black liquor properties vary greatly from mill to mill. Some mills' black liquor can inhibit enzyme performance due to the presence of highly oxidizing compounds. This effect differs significantly among enzymes and should always be checked before proceeding with fullscale enzyme use. It is important to note that it is not necessary to wash the pulp after enzyme treatment (before chlorination) to achieve the enhanced bleaching. Identical enzyme benefits with and without a post-enzyme washing have been obtained. The bleaching sequence and brightness target influence the enzyme's benefits to the mill. The enzyme benefit is greater at higher brightness targets, especially near the brightness ceiling, and lower at lower brightness targets.

Valls et al. (2010) used two new bacterial xylanases from families 11 and 5 to obtain modified fibers with high cellulose content. When these xylanases were applied separately or simultaneously in a complete elemental chlorine free (ECF) bleaching sequence, both xylanases were found to improve delignification and bleaching during the sequence, while a synergistic effect of the enzymes was observed on several pulp and paper properties. The xylanases enhanced the release of xylo-oligosaccharides branched with hexeneuronic acids (HexA), producing fibers with a reduced HexA and xylose content. On the other hand, these effects were found to be dependent on the xylanase used, with the family 11 enzyme being more efficient than the family 5 xylanase. Effluent properties were affected by the enzymatic sequences, due to the dissolution of lignin and xylooligosaccharides, while some changes in the fiber morphology were also produced without affecting the final paper strength properties.

Xylanase pretreatment of pulps prior to bleach plant reduces bleach chemical requirements and permits higher brightness to be reached. The reduction in chemical charges can translate into significant cost savings when high levels of chlorine dioxide and hydrogen peroxide are being used. A reduction in the use of chlorine chemicals clearly reduces the formation and release of chlorinated organic compounds in the effluents and the pulps themselves. The ability of xylanases to activate pulps and increase the effectiveness of the bleaching chemicals may allow new bleaching technologies to become more effective. This means that for expensive chlorine-free alternatives, such as ozone and hydrogen peroxide, xylanase pretreatment may eventually permit them to become cost-effective. Traditional bleaching technologies also stand to benefit from xylanase treatments. Xylanases are easily applied and require essentially no capital expenditure. Because chlorine dioxide charges can be reduced, xylanase may help eliminate the need for increased chlorine dioxide generation capacity. Similarly, the installation of expensive oxygen delignification facilities may be avoided. The benefit of a xylanase bleach boosting stage can also be taken to shift the degree of substitution toward higher chlorine dioxide levels while maintaining the total dosage of active chlorine. Use of high chlorine dioxide substitution dramatically reduces the formation of AOX.

In totally chlorine free-bleaching sequences, the addition of enzymes increases the final brightness value, which is a key parameter in marketing chlorine-free pulp. In addition, savings in TCF bleaching are important with respect both to costs and to the strength properties of the pulp. The production of TCF pulp has increased dramatically during recent years. Several alternative new bleaching techniques based on various chemicals, such as oxygen, ozone, peroxide, and peroxyacids have been developed. In addition, an oxygen delignification stage has already been installed at many kraft mills. In the bleaching sequences in which only oxygen-based chemicals are used, xylanase pretreatment is generally applied after oxygen delignification to improve the otherwise lower brightness of the pulp or to decrease bleaching costs. The TCF sequences usually also contain a chelating step in which the amount of interfering metal ions in pulp is decreased. It has been observed that the order of metal removal (Q) and enzymatic stages (X) is important for an optimal result. When aiming at the maximal benefit of enzymatic treatment in pulp bleaching, the enzyme stage must be carried out prior to or simultaneously with the chelating stage. In fact, the neutral pH of enzyme treatment is optimal in many cases for the chelation of magnesium, iron, and manganese ions that must be removed before bleaching with hydrogen peroxide. The TCF technologies applied today are usually based on the bleaching of oxygen-delignified pulps with enzymes and hydrogen peroxide.

A survey of mill usage of xylanase revealed that the mills have spent most of their efforts in decreasing AOX (by decreasing chlorine usage), followed closely by meeting customer demands (which in many cases was decreasing chlorine usage), and eliminating chlorine gas. These objectives were followed in effort by decreasing off-grade pulp, decreasing BOD, and cutting costs. The least effort was devoted to increasing throughput, eliminating dioxin, and converting to TCF (Tolan et al., 1996). The most widely-reported benefit of enzyme treatment is a savings in bleaching chemicals. The chemical savings was 8% to 15% with an average of 11% of the total chemical across the bleach plant. The other widespread benefits were in improved effluent, including decreases in AOX of 12% to 25%, decreases in effluent color, and other improvements to the effluent. Other benefits of enzyme treatment reported increased bleached brightness (1 point gain), tear strength (5% gain), and pulp throughput (10% increase). Xylanase enzymes can cut bleaching related energy usage by 40%. This would result in carbon dioxide emissions savings of between 155,000 and 270,000 tons annually in the European paper industry.

The most common problems with xylanase treatment cited in a mill survey have been corrosion of equipment and maintaining the brownstock residence time. Sulfuric acid corrosion of mildsteel has been encountered in several mills. The brownstock residence time must be maintained for as long as possible, but usually at least 1 to 2 hours to obtain the maximum benefits of the enzyme treatment. This sometimes means that the mills must maintain the storage tower nearly full, which curtails its ability to act as a buffer between the pulping mill and the bleach plant. Other problems reported with enzyme treatment included difficulties in application and in bleach plant control. These relate to subtle actions of enzymes, which are not easily observed on-line or in rapid testing. A decreased tear strength and pitch formation were also reported in some mills.

Xylanase-aided bleaching has been identified as a future technology. The development is focusing on improved enzyme properties and improved enzyme performance. Improved properties include higher pH and temperature tolerance of the enzymes, to make the enzyme treatment operations more compatible with existing mill operations. Improved enzyme performance is being approached by tailoring the enzyme action more closely to the hemicellulose structure of the pulp, to result in a greater bleaching benefit or a higher pulp yield.

8.1.2 Fiber Modification

The enzymatic modification of fibers aims at the decreased energy consumption in the production of thermomechanical pulps and the increased beatability of chemical pulps or the improvement of fiber properties. Xylanase action has been found to produce pulp fibers with properties similar to those of slightly beaten pulps. Laboratory and process-scale studies conducted with a mixture of xylanase and cellulase enzymes with different pulps—hardwood Kraft pulp, long fiber fraction of bamboo pulp, OCC and mixed pulp containing NDLKC, and long fiber fraction of bamboo pulp—showed 18–45% reduction in refining energy and no adverse effect on the physical strength properties of pulps. Enzyme-aided refining is economically attractive and is now a commercial reality (Bajpai et al., 2005a, 2006; Gill, 2008; Loosvelt, 2009). It is easy to integrate into the present production processes and causes no problem in normal operations. The main benefits of enzymatic refining include reduction in electrical energy for refining, reduction in steam consumption, and reduction in backwater consistency. These benefits can be converted into increased machine speed, improved machine performance, reduction in retention aids, enhanced formation of paper debottlenecking of refiner capacity, and the possibility of utilizing difficult to refine pulp. Potential benefits of enzymatic refining include ease in operation of backwater clarification, possibility of reduction in pitch problems, improved biodegradability of machine effluent, and reduction in greenhouse gas emissions.

Water removal on the paper machine has been shown to improve as a result of limited hydrolysis of the fibers in recycled paper. A mixture of xylanase and cellulase enzymes at low concentrations has been found to markedly increase the freeness of recycled fibers without reducing the yield (Bajpai et al., 2004). The lower the initial freeness, the greater the gain following treatment. Freeness shows a rapid initial increase with over half of the observed effect occurring in the first 30 minutes. A relatively small amount of enzyme is required. Although the initial effects are largely beneficial, extending the reaction time with large concentrations of enzyme is detrimental. During enzyme treatment of OCC fibers with commercial xylanase, the greatest improvements in drainage rate were obtained when xylanase was applied to fiber after beating (Bobu et al., 2003). The greatest improvements in relative bonded area and wet fiber flexibility were recorded with xylanase applied before beating. Xylanase treatment induces swelling of the fibers, which has a strong influence on any subsequent processing.

8.1.3 Deinking of Waste Paper

Xylanase enzymes in combination with cellulases are being used for the deinking of waste paper. (Bajpai and Bajpai, 1998; Bajpai, 2006, a, b, 2012). This method has proven effective and economical on both a laboratory and industrial scale. Cellulases and xylanases exhibit significant effect on the enzymatic deinking of old newsprint (ONP), improving deinking efficiency and fiber modification (Wang and Kim, 2005). The most promising implication of high deinking efficiency from enzyme enhanced deinking is that the dewatering and dispersion steps and the subsequent flotation and washing may not be essential. This should save capital expenses in construction of deinking plants while also reducing consumption of electrical energy for dewatering and dispersion. The requirement of bleach chemicals is usually lower for enzymatic deinking than for conventional chemical deinking. Reduced chemical use would result in lower waste treatment costs while reducing the impact on the environment. Lower bleaching costs and less pollution can also be anticipated, since enzymatically deinked pulps have proved to be easier to bleach and require less chemicals than pulps deinked by conventional methods. Enzymatically deinked pulp also displays improved drainage, superior physical properties, higher brightness, and lower residual ink compared with chemically deinked recycled pulps. Improved drainage results in faster machine speed, which yields significant savings in energy and thus overall cost. In addition, the use of recycled fiber reduces the need for virgin pulp. This results in great savings in the energy required for pulping, bleaching, and refining,which will also eventually reduce pollution problems.

8.1.4 Production of Dissolving Pulp

Dissolving pulp is a high-grade cellulose pulp, with low contents of hemicellulose, lignin, and resin. This pulp has special properties, such as a high level of brightness and uniform molecular-weight distribution. Dissolving-grade pulps are commonly used for the production of cellulose derivatives and regenerated cellulose (Bajpai, 1997, 1999, 2012; Paice and Zhang, 2005). To obtain products of high quality, these pulps must fulfill certain requirements, such as high cellulose content, low hemicellulose content, a uniform molecular weight distribution, and high cellulose reactivity. Xylan-degrading enzymes have been explored for selective removal of pentosans in preparing dissolving grade pulp (Christov and Prior, 1993, 1994, 1996; Christov et al., 1995; Bajpai and Bajpai, 2001b; Bajpai et al., 2005b). The complete removal of residual hemicellulose seems unachievable due to the modification of the substrate or to structural barrier. It appears that pentosans in the bleached pulp are well-shielded by other pulp components and are therefore not susceptible to enzymatic attack. Even with very high enzyme loadings and prolonged incubation periods, xylan hydrolysis is limited. The wood species and the methods of their pulping, the accessibility of pentosans and their quantity in pulp, the penetration capabilities and substrate specificity of the enzymes, the inhibitory action of bleaching chemicals, and the linkage of xylan to lignin and cellulose by covalent and hydrogen bonds, respectively, may be the factors contributing to the difficulties in removing xylan from bleached pulp (Minor, 1986; Gamerith and Strutzenberger, 1992; Henriksson et al., 2005). Xylanase treatment of unbleached pulp appears to be more effective because of the presence at this stage of more hemicellulose accessible for enzymatic degradation. Alkaline extraction in conjunction with enzyme treatment leads to some improvement of the pulp characteristics.

8.1.5 Shive Removal

Cleanliness is one of the very important quality parameters of a bleached chemical pulp (Smook, 1992; Reeve, 1996a, b). The cleanliness of bleached pulp is affected by coarse particles derived from the cellular tissue of the tree and by extraneous particles. The former group consists of shives, knots, and bark specks. The latter group consists of pitch, fungus hyphae, rust strains, lime, and sand. A shive is a particle or fiber bundle large enough, or in enough quantity, to produce a paper and board quality or productivity. Xylanases are also effective in the removal of shives (small bundles of fibers that have not been separated into individual fibers during the pulping process). Shives appear as splinters that are darker than the pulp. One of the most important quality criteria for bleached kraft pulp is shive count. By treating the brownstock with xyanases, mills can substantially increase the degree of shive removal in the subsequent bleaching. Xylanase enzymes have also been found to be effective in removing shives (Bajpai, 1997, 1999, 2009; Tolan et al., 1994). Xylanase enzymes can increase the degree of shive removal by 50% in subsequent bleaching. At a given bleached brightness, xylanase treatment results in a lower shive count. Xylanase treatment, therefore, helps to remove shives from the pulp beyond the associated gain in the brightness. Removal of shives and ease of pulp bleaching by the use of xylanases also help in reducing the energy requirement (Tolan et al., 1994).

8.1.6 Removal of Bark

The removal of tree bark is the first step in all processing of wood (Smook, 1992). This step consumes a substantial amount of energy. Extensive debarking is needed for high-quality mechanical and chemical pulps because even a small amount of bark residue causes the darkening of the product. In addition to its high energy demand, complete debarking leads to the loss of raw material due to prolonged treatment in the mechanical drums. The border between the wood and bark is the cambium, which consists of only one layer of cells. This living cell layer produces xylem cells toward the inside of the stem and phloem cells toward the outside. In all the wood species, the common characteristics of the cambium include a high content of pectins and the absence or low content of lignin. The content of pectins in cambium cells varies among the wood species, but may be as high as 40% dry weight. The content of pectic and hemicellulosic compounds is very high in the phloem. Pectinases are found to be the main enzymes in the process, but xylanases may also play a role because of the high hemicellulose content in the phloem of the cambium. Enzymatic treatments cause significant decreases in energy consumption during debarking (Grant, 1992, 1993, 1994; Bajpai, 1997, 2006b, 2009; Hakala and Pursula, 2007). The energy consumed in debarking is decreased as much as 80% after pretreatment with pectinolytic enzymes. The enzymatic treatment also results in substantial savings in raw material. Enzymes may be able not only to increase existing debarking capacity, thus saving capital investment, but may also be available as an aid to be used when debarking is difficult. Finnish researchers (Ratto et al., 1993; Viikari et al., 1989, 1991) used debarking enzymes, specifically for the hydrolysis of the cambium and phloem layer, from *Aspergillus niger*. A clear dependence was found between the polygalacturonase activity in the enzyme preparation and reduced energy consumption in debarking. In addition to polygalacturonase, the enzyme mixture produced by *A. niger* also contained other pectolytic and hemicellulolytic activities. The amount of energy needed for removal of bark was found to decrease to 20% of the reference value.

8.1.7 Retting of Flax Fibers

Flax provides a variety of important industrial products, such as textiles, oilseed, and paper/pulp. Fibers from flax are the oldest textile known. It has been considered a prime source of natural fibers to replace glass in composites. Flax is also the source of linseed oil for industrial uses and nutritional flaxseeds. Fiber flax produced for textiles is grown under precise conditions to optimize fiber quality and is harvested prior to full seed maturity. Fibers are obtained from flax stems by the process of retting. Two methods employed for retting flax at commercial levels using pectinolytic microorganisms are water- and dew-retting. Water-retting traditionally depends upon anaerobic bacteria that live in lakes, rivers, ponds, and vats to produce pectinases and other enzymes to ret flax. The stench from anaerobic fermentation of the plants, extensive pollution of waterways, high drying costs, and putrid odor of resulting fibers resulted in a move away from anaerobic water-retting in the mid 20th century to dew-retting. Dew-retting is the result of colonization and partial plant degradation by plant-degrading, aerobic fungi of flax stems, which are harvested and laid out in swaths in fields. The highest quality linen fibers are produced using dew-retting, but concern exists within this industry about low and inconsistent quality. Enzymes have the potential to provide an improved method to ret flax for textile fibers. Enzymatic retting produces high and consistent quality fibers of staple length for blending with cotton and other fibers (Sharma, 1987; Gillespie et al., 1990; Van Sumere, 1992; Bajpai, 1997, 1999; Akin et al., 1997, 2000; Akin, 1998; Ebbelaar et al., 2001; Hoondal et al., 2002; Kenealy and Jeffries, 2003; Antonov et al. 2005; Foulk et al., 2008). Xylanase enzymes have been also used in processing plant-fiber sources, such as flax and hemp. Pectinases are believed to play the main role in this process, but xylanases may also be involved. The replacement of slow natural retting by treatment with artificial mixtures of enzymes could become a new fiber-liberation technology.

8.2 FOOD

Xylanases are widely applied in the food industry (Polizeli et al., 2005; Butt et al., 2008; Harris and Ramalingam, 2010). Xylanases together with α-amylase, malting amylase, glucose oxidase, and proteases may be employed in bread making. Xylanases are vital for bread making and are one of the most established baking enzyme applications. A fundamental function of xylanases is the degradation of water-unextractable arabinoxylan (a hemicellulose found in wheat) into water-extractable arabinoxylan. This enables redistribution of water in the dough, which, in turn, improves gluten network formation in the dough. The desired outcome is a dry dough balanced correctly in terms of the right softness, extensibility, and elasticity.

Enzymatic hydrolysis of nonstarch polysaccharides leads to the improvement of rheological properties of dough, bread specific volume, and crumb firmness. The xylanases, like the other hemicellulases, break down the hemicellulose in wheat flour, helping in the redistribution of water and leaving the dough softer and easier to knead. During the bread baking process, they delay crumb formation, allowing the dough to grow. With the use of xylanases, there has been an increase in bread volumes, greater absorption of water, and improved resistance to fermentation (Maat et al., 1992; Harbak and Thygesen, 2002; Camacho and Aguilar, 2003). Also, a larger amount of arabinoxylooligosaccharides in bread would be beneficial to health (Polizeli et al., 2005). A particular endo-1,4-ß xylanase produced by *Aspergillus niger* has been identified that is very effective in increasing the specific volume of breads without giving rise to a negative side effect on dough handling, similar to that observed with xylanases derived from other bacterial or fungal sources (Monfort et al., 1997; Sorensen et al., 2001; Wang et al., 2004; Butt et al., 2008). The addition of *A. niger* var. *awamori* endoxylanase to dough enhanced the specific volume, color, and crumb structure of bread upon baking when compared to the control without adding the enzymes (Van Gorcom et al., 2003).

Xylanases improve dough characteristics and bread quality leading to improved dough flexibility, machinability, stability, loaf volume, and crumb structure (Baillet et al., 2003; Guy and Sarabjit, 2003). Many enzymes, such as proteases, xylanases, and cellulases, improve the strength of the gluten network and, therefore, improve the quality of bakery products (Gray and BeMiller, 2003). The enzymatic hydrolysis of pentosans by hemicellulases or pentosanases at the optimal level improves dough properties resulting in greater uniformity in quality characteristics (Rouau et al., 1994). Xylanases make dough more tolerant to different flour quality parameters and variations in processing methods. They also make dough soft, reduce the sheeting work requirements, and significantly increase the volume of leavened pan bread (Dervilly et al., 2002; Harbak and Thygesen, 2002). Xylanase, along with protease, lipase, and α-amylase, is very effective for obtaining bread with higher specific volume in a microwave oven, as compared to the bread with no enzyme added. The texture profile analysis was greatly modified by xylanases and the firmness of bread crumb was reduced (Mathewson, 2000; Ozmutlu et al., 2001).The positive effect of xylanase on bread volume is due to the redistribution of water from the pentosan phase to the gluten phase. The increase in the volume of the gluten fraction increases its extensibility, which results in better oven spring (Maat et al., 1992). The improving effect of pentosanases on bread volume may be associated with better gas retention during proofing, probably due to the action of the enzyme in reducing the viscosity of the gelling starch and allowing greater and longer expansion in the oven before enzyme inhibition and protein denaturation (Martinez-Anaya and Jimenez, 1997).

Novozymes has introduced Panzea, from a new generation of xylanases, claimed to be the first baking xylanase to significantly combine superior volume performance and desired texture and appearance with a dry, balanced dough—all in one product. It performs at very low dosage across a broad range of bread products, baking conditions, and flour types. Unlike most xylanases, Panzea is naturally uninhibited by xylanase inhibitors commonly found in wheat. This means it delivers robust performance at very low dosages across different types of flours, so it's both more efficient and more versatile than what is already on the market. Bread improver companies and flour mills often have to use a variety of xylanases to find exactly the right combination to achieve the desired baking result. But with Panzea, it is now possible to meet a wider variety of needs. Panzea, produced in *Bacillus Licheniformis*, has been internally and externally documented to achieve the desired loaf volume, crumb structure, and dough characteristics across a very broad range of baking procedures and flours, at low dosages (Novozymes, 2013) .

On the comparison of the abilities of different xylanases isolated from *Aspergillus oryzae, Humicola insolens*, and *Trichoderma reesei* to improve the quality of bread made from wheat flour, it was shown that the most effective is xylanase from *A. oryzae* (Basinskiene et al., 2006).The specific volume of bread increased by 8−13% and crumb firmness decreased by 15−24% compared to bread without xylanase. However, maximum antistaling effect was observed with xylanase from *T. reesei*. Jiang et al. (2005) isolated a xylanase from thermophilic bacteria *Thermomyces lanuginosus* CAU44 and showed its application in bread making. Laurikainen et al. (1998) reported an increase in the softening of wheat dough from 90 BU in the control to 170 BU in xylanase-supplemented dough when *Tricoderma* culture filtrate enriched in endo-1,4-β-xylanase was added. Martinez-Anaya and Jimenez (1997) reported that the use of starch and nonstarch hydrolyzing enzymes released free water and changed the soluble fraction of dough. Redgwell et al. (2001) reported a constant decrease in viscosity of wheat flour batter. They showed that the change in viscosity was caused by the action of xylanase on the crude enzyme preparation. Jiang et al. (2005) also observed that the specific volume of wheat bread improved using xylanase B from *Thermotoga maritima*. Pescado-Piedra et al. (2009) studied the effect of the addition of several enzymes—glucose oxidase, peroxidase, and xylanase—on the rheological parameters of dough and the quality of the bread. Addition of peroxidase and xylanase was found to increase water absorption, while the addition of glucose oxidase had no effect. Crumb characteristics were improved by xylanase. High concentration of glucose oxidase or xylanase decreased the stability (Garg et al., 2010).

In biscuit-making, xylanase is used for making cream crackers lighter and also for improving the texture, palatability, and uniformity of the wafers (Polizeli et al., 2005).

Xylanases, in conjunction with cellulases, amylases, and pectinases, lead to an improved yield of juice by means of liquefaction of fruit and vegetables; stabilization of the fruit pulp; increased recovery of aromas, essential oils, vitamins, mineral salts, edible dyes, and pigments; reduction of viscosity; and hydrolysis of substances that hinder the physical or chemical clearing of the juice, or that may cause cloudiness in the concentrate (Polizeli et al., 2005; Biely, 1985). Xylanase reduces the viscosity of the brewing liquid, improving its filterability (Garg et al., 2010). Xylanase, in combination with endoglucanase, takes part in the hydrolysis of arabinoxylan and starch, separating and isolating the gluten from the starch in the wheat flour. This enzyme is also used in coffee bean mucilage (Wong and Saddler, 1993; Wong et al., 1988).

The main desirable properties for xylanases for use in the food industry are high stability and optimum activity at an acid pH. With advances in the techniques of molecular biology, other uses of xylanases are being discovered (Polizeli et al., 2005). A recombinant yeast of wine was constructed with the gene for xylanase of *Aspergillus nidulans*, xlnA, resulting in a wine with a more pronounced aroma in comparison to the conventional (Ganga et al., 1999). During the manufacture of beer, the cellular wall of the barley is hydrolyzed, releasing long chains of arabinoxylans, which increase the beer's viscosity rendering it "muddy" in appearance. Thus, xylanases are used to hydrolyze arabinoxylans to lower oligosaccharides diminishing the beer's viscosity and consequently eliminating its muddy aspect (Debyser et al., 1997; Dervilly et al., 2002). α-Larabinofuranosidase and β-D-glucopyranosidase have been employed in food processing for aromatizing musts, wines, and fruit juices (Spagna et al., 1998). The hydrolysis products of xylan–xylose and oligosaccharides–have possible applications in food industry. These can be used as thickeners, as fat substitutes and as antifreeze food additives.

8.3 FEED

Xylanases are used in the pretreatment of forage crops to improve the digestibility of ruminant feeds and to facilitate composting along with glucanases, pectinases, cellulases, proteases, amylases, phytase, galactosidases, and lipases (Gilbert and Hazlewood, 1993; Polizeli et al., 2005; Juturu and Wu, 2012; Sharma and Kumar, 2013). These enzymes break down arabinoxylans in the ingredients of the feed, reducing the viscosity of the raw material (Twomey et al., 2003). The arabinoxylan found in the cell walls of grains has an anti-nutrient effect on poultry. When such components are present in soluble form, they may raise the viscosity of the ingested feed, interfering with the mobility and absorption of other components. If xylanase is added to feed containing maize and sorghum, both of which are low viscosity foods, it may improve the digestion of nutrients in the initial part of the digestive tract, resulting in a better use of energy. The joint action of the rest of the enzymes listed produces a more digestible food mixture. Young fowl and swine produce endogenous enzymes in smaller quantities compared to adults, so that food supplements containing exogenous enzymes should improve their effectiveness for livestock. Furthermore, this type of diet reduces unwanted residues in the excreta (phosphorus, nitrogen, copper, and zinc), an effect that could reduce environmental contamination (Polizeli et al., 2005).

Café et al. (2006) gave nutritionally rich diets, with or without the addition of 0.1% Avizyme 1500 (xylanase, protease, and amylase), to poultry birds. Birds fed on the diets supplemented with Avizyme exhibited significantly higher body weights, less mortality, and a greater amount of net energy from their diets as compared to the control group. Babalola et al. (2006) observed improved apparent nitrogen and fiber absorption as well as feed transit time by the application of xylanase in poultry feed. Moreover, the enzyme addition in boiled castor seed meal (up to 150 g/kg) was found to be acceptable and showed no adverse affect on growth performance or blood constituents.

Substantial amounts of barley, wheat, rye, and sunflower meal can only be used in broiler diets if they are supplemented with enzymes. The addition of enzymes ensures the maximum utilization of nutrients trapped inside the plant cells, while also reducing the viscosity created by nonstarch polysaccharides (NSP) in the bird's digestive tract. For ruminants, alteration in milk fat content and improvement in the productivity of beef cattle has been demonstrated by adding the enzymes beta-glucanase and xylanase to animal feed. Additionally, by including these enzymes to the total mixed feed ration or to corn silage or alfalfa hay, improved animal health has been demonstrated—especially in young animals. In pig feed, increasing levels of xylanase supplementation, up to 600 g of the commercial enzyme per ton of feed (containing 20−35% barley and 30−35% wheat), has resulted in improvement of performance, as measured by feed conversion ratio and live weight gain. Pig feeds normally contain more by-products than poultry feeds because of the overall higher digestive efficiency of the pig compared with poultry. These by-products can also be upgraded by enzyme supplementation and, in some instances, can offer higher potentials than the materials from which they are derived. The presence of noncereal feed raw materials in the complex diet (e.g., rapeseed meal, sugar beet pulp, sunflower meal) has been found to diminish the enzyme effect, as the substrates present were not degradable by the xylanase used. This implies that attention must be paid to the raw material composition of pig diets if the maximum enzyme effect is to be achieved.

Xylanases also convert hemicelluloses to sugars and, thereafter, nutrients trapped within the cell walls are released and chickens get sufficient energy from lesser feed. The barn is cleaner due to more thoroughly digested feed and the chicken waste is drier and less sticky. In addition, chicken eggs are cleaner due to a dry laying area (Garg et al., 2010). Mathlouthi et al. (2002) showed that feeding a rye-based diet to chickens reduces villus capacity for nutrient absorption and bile acid capacity for fat solubilization and emulsification.

It is known that xylanase and cellulase treatment of forages produces a better quality silage that improves the subsequent rate of plant cell wall digestion by ruminants. A formulation containing both xylanases and *Lactobacillus* was sprayed over silage. Xylanases present in the formulation selectively depolymerized hemicellulose-producing xylose, which was fermented to lactic acid by *Lactobacillus* thereby increasing the stability, digestibility, and nutritive values of the silage for digestion by cattle (Evans et al., 1995a, b).

There is a considerable amount of sugar sequestered in the xylan of plant biomass. As a result of xylanase treatment, there is increased nutritive sugar, which is useful for digestion in cows and other ruminants. It is also known that xylanase produces compounds that are the nutritive source for much ruminal microflora (Garg et al., 2010).

8.4 TEXTILES

A number of reports are available that cite the use of cellulases and pectinases for bioprocessing of fabric. There are also a few reports on xylanases for the purpose of desizing and scouring (Csiszar et al., 2001; Losonczi et al., 2005; Dhiman et al., 2008a). The xylanolytic complex can be used in the textile industry to process plant fibers, such as hessian or linen. For this purpose, the xylanase should be free of cellulolytic enzymes. One process consists of incubating dried ramee (China grass) stems with xylanase to liberate the long cellulose fibers intact. After using this method, there is no need to use the strong bleaching step, as the lignin does not undergo oxidation, which would lead to darkening of the fibers (Prade, 1995; Brühlmann et al., 2000; Csiszar et al., 2001). Processing of the fabric includes desizing, scouring, and bleaching (Karmakar, 1999; Rouette, 2001). Sizing materials protect the fabric against abrasive forces during weaving. Desizing is carried out to remove the adhesive material in order to render the fabric more accessible to the subsequent stages of the processing. Conventionally, it is carried out at higher temperatures with strong oxidizing agents in alkaline solutions. After desizing, the fabric needs to be scoured to remove inhibitory materials for its efficient finishing, wetting, and dying (Harris et al., 1998). Conventional scouring is a chemical-intensive, nonoptimal process and attacks, nonspecifically, cellulosic material of fiber, which in turn causes strength loss. The advantage associated with enzymatic pretreatment is the highly specific action of the enzyme. Dhiman et al. (2008b) reported that xylanase specifically acted on the hemicellulosic impurities and efficiently caused their removal. Enzymatic treatment did not cause any strength loss of the fiber. The greatest hurdle in the commercialization of the enzymatic method is offered by the presence of seed coat fragments attached to the fibers and linters (Verschraege, 1989). Enzymatic pretreatment with xylanase leads to the partial hydrolysis of these seed coat fragments, thereby making them more accessible to the chemicals during the later stages of bleaching and finishing.

Battan et al. (2012) studied the desizing of cotton and micropoly fabrics using thermostable xylanase from *Bacillus pumilus* ASH. Micropoly fabric showed better desizing than cotton under the same conditions. Violet scale readings from the TEGEWA test after enzymatic desizing for 90 min at pH 7.0 and 60°C showed the readings falling in the range of 4–5, indicating good desizing efficiency. During bioscouring, the weight loss values and liberation of reducing sugars were highest when EDTA was used along with xylanase. The weight loss value of 1.5% was observed for dry cotton fabric after 1 h in the case of an agitated system at pH 7.0 and at an optimal enzyme dosage of 5 IU/g. The weight loss values and the liberation of reducing sugars were higher in the case of cotton fabrics. The wetting time of fabrics was lowered significantly after 60 min of bioscouring using xylanase. An increase in temperature or concentration of surfactant led to a further reduction in the wetting time. The whiteness values of fabrics after bioscouring were 0.9% higher than the chemically scoured fabrics, indicating good efficacy of xylanase during the scouring process.

Garg et al. (2010) evaluated alkalo-thermostable xylanase from *Bacillus pumilus* ASH for bioscouring of jute fabric. An enzyme dose of 5 IU/g of oven-dried jute fabric resulted in release of more reducing sugar and weight loss as compared to control when incubated at 55°C. An incubation time of 120 min was sufficient to increase the whiteness and brightness of fabric up to 3.93 and 10.19%, respectively, and also decreased the yellowness by 5.57%. The addition of chelating and wetting agents greatly enhanced the fabric properties. Bioscouring of jute fabric with xylanase enzymes along with EDTA and Tween-20 resulted in an increase of 9.63, 4.28, and 10.71% of reducing sugars, whiteness, and brightness, respectively, when compared to conventional process. Bleaching of bioscoured jute fabric, further improved the various properties like tenacity, brightness, yellowness, and whiteness.

8.5 PHARMACEUTICALS AND CHEMICALS

Xylanase and xylan are little used in the pharmaceutical industry. Xylanases are sometimes added in combination with a complex of enzymes (hemicellulases, proteases, and others) as a dietary supplement or to treat poor digestion, but few medicinal products can be found with this formulation.

Xylanase, together with other hydrolytic enzymes, can be used for the generation of biological fuels, such as ethanol, from lignocellulosic biomass (Olsson and Hahn-Hägerdal, 1996; Screenath and Jeffries, 2000; Beg et al., 2001; Bajpai, 2012, 2013b). The process of ethanol production from lignocellulosic biomass includes the delignification of plant biomass and the hydrolysis of cellulose and hemicellulose to monosaccharides (Beg et al., 2001). The hydrolysis process can be performed by treatment with acids at high temperatures or by enzyme action. The acidic hydrolysis requires a significant energy consumption and acid-resisting equipment, which makes the process more expensive. However, enzymatic hydrolysis does not have these disadvantages. Because of the complex composition of lignocellulosic biomass, the synergistic action of several enzymes with respect to endoglucanases, β-glucosidases, endo-1,4-β-xylanases, and β- xylosidases is required for complete hydrolysis (deVries and Visser, 2001). In some cases, endo-1,4-β-xylanase has been reported to be a bifunctional enzyme having endo-1,4-β-xylanase as well as cellulase activity. Bi-functionality of endo-1,4-β-xylanase could result in a more efficient and cheaper saccharification process of agricultural residues and both municipal and industrial wastes used for bio-ethanol production, as it can degrade both cellulose and xylan residues. Saccharification of the cellulose and hemicellulose in biomass results in sugar-rich liquid streams useful for the production of a variety of value-added products including ethanol, furfural, and various functional biopolymers (Fuller et al., 1995). An increased possibility of fermentation of both hexose and pentose sugars present in lignocelluloses into methanol has also been reported (Senn and Pieper, 2001). Simultaneous saccharification and fermentation is an alternative process, in which both hydrolytic enzymes and fermentative microorganisms are present in the reaction (Pérez et al., 2002; Chandrakant and Bisaria, 2008). However, enzymatic hydrolysis is still a major cost factor in the conversion of lignocellulosic raw materials to ethanol (Viikari et al., 2012).

Xtreme Xylanase, the most thermal- and acid-stable xylanase enzyme from *Alicyclobacillus acidocaldarius* discovered by INL researchers, is capable of efficiently converting the hemicelluloses and cellulose components of biomass into energy-rich sugars. These sugars are the building blocks used in place of petroleum to make fuels and high-value chemicals (US Department of Energy, 2012). By virtue of its extreme acidity and thermostability, Xtreme Xylanase will revolutionize biorefineries in several ways:

• Reduce or eliminate pretreatment costs,
• Improve fermentation efficiency,
• Improve downstream economics,
• Streamline biorefinery flowsheets, and
• Complement consolidated bioprocessing strategies.

Xylanase, in synergism with several other enzymes such as mannanase, ligninase, xylosidase, glucanase, and glucosidase, can be used for the production of xylitol, from lignocellulosic biomass (Kuhad and Singh, 1993). Xylitol is a polyalcohol with a sweetening power comparable to that of sucrose (Parajó et al., 1998). It is a noncariogenic sweetener, suitable for diabetic and obese individuals and recommended for the prevention of osteoporosis and respiratory infections, lipid metabolism disorder, and kidney and parenteral lesions. A variety of commercial products containing xylitol, such as chewing gum, can be found on the market. It has attracted much attention because of its applications in foods and confectionery (Saha and Bothast, 1997, 1999). The bulk of xylitol produced is consumed in various food products such as chewing gum, candy, soft drinks, and ice cream. It gives a pleasant cool and fresh sensation due to its high negative heat of solution. The recovery of xylitol from the xylem fraction is about 50−60% or 8−15% of the raw material employed (Winkelhausen and Kuzmanova, 1998). The value depends on the xylem content of the raw material.

Although the enzymatic hydrolysis of xylan is a promising method of obtaining β-D-xylopyranosyl units, at present commercial xylitol is produced on a large scale by chemical catalysis. This is considered a high-cost process, mainly because the xylose has to be purified initially in several steps. Besides this, the chemical reactions often produce by-products toxic to fermentation; indeed, in the decomposition of lignocellulosic material, besides the liberation of sugars, products may be formed that are derived from the degradation of glucose (hydroxymethylfurfural), xylose (furfural), and lignin (aromatic and phenolic compounds and aldehydes). Substances liberated from the lignocellulose structure, such as acetic acid and extracted material (e.g., terpenes and their derivatives, tropolones, and phenolic compounds such as flavonoids, stilbenes, quinones, lignans, and tannins), or from the equipment (iron, chromium, nickel, and copper), can be powerful inhibitors of microbial activity. The development of a more appropriate technology for xylitol production has generated great hope of its wider use in the food, pharmaceutical, and odontological industries.

A recent and exciting application of xylanases is the production of xylo-oligosaccharides (XOs) (Sharma and Kumar, 2013); these are newly developed functional oligosaccharides that feature many beneficial biomedical and health effects, such as the stimulation of human intestinal *Bifidobacteria* growth (Yang et al., 2005). Currently, XOs are produced mainly by enzymatic hydrolysis due to high specificity, negligible substrate loss, and side-product generation (Tan et al., 2008). A research team at a Belgian University has isolated a cold-active family 8 xylanase. This enzyme offers potential applications for the preparation of xylo-oligosaccharides for use as prebiotics in either food or nonfood applications. If produced cost effectively, XOS could play a significant part of the quickly enlarging prebiotics markets. Other applications are: in the preparation of soluble fiber enriched products, as well as in the pulp and paper, textile, food, and biofuel industries. The research team is looking for collaboration with companies for joint development, testing, and licensing. It has been estimated that by 2015, the prebiotics markets will total 880 M€/a in Europe and 170 M€/a in the US. Currently xylo-oligosaccharides are sold in China and Japan for prebiotic purposes in commercial volumes.

Xylan degradation requires the interaction of several enzymatic activities, including endo-1,4-β-xylanase, xylosidase, and arabinosidase. The end-products of this degradation are xylose, arabinose, and methylglucuronic acid containing oligosaccharides. Xylooligosaccharides are sugar oligomers showing potential for practical applications in a variety of fields, including pharmaceuticals, feed formulations, agricultural purposes, and food applications (Vazquez et al., 2000). As additives for functional foods, XOs have prebiotic action, showing positive biological effects, such as improvement in the intestinal function by increasing the number of healthy *Bifidobacteria* (Rycroft et al., 2001; Fooks and Gibson, 2002; Izumi and Kojo, 2003). These xylooligosaccharides—if used as dietary supplements—may be beneficial to gastrointestinal health and may reduce the risk of colon cancer (Whitehead and Cotta, 2001). As food ingredients, XOs have an acceptable odor and are non-carcinogenic (Kazumitsu et al., 1987; Kazumitsu et al., 1997) and have a low-calorie value, allowing their utilization in anti-obesity diets (Toshio et al., 1990; Taeko et al., 1998). In food processing, XOs show advantages over insulin in terms of resistance to both acids and heat, allowing their utilization in low-pH juices and carbonated drinks (Modler, 1994).

REFERENCES

Akin, D., 1998. Enzyme retting of flax for linen fibers: recent developments. Book of Papers American Association Textile Chemists and Colorists. American Association Textile Chemists and Colorists, Research Triangle Park (NC), pp. 273–280.

Akin, D., Morrison, W., Gamble, G., Rigsby, L., 1997. Effect of retting enzymes on the structure and composition of flax cell walls. Textile Res. J. 67 (4), 279–287.

Akin, D., Dodd, R., Perkins, W., Henriksson, G., Eriksson, K., 2000. Spray enzymatic retting: a new method for processing flax fibers. Textile Res. J. 70 (6), 486–494.

Allison, R.W., Clark, T.A., Wrathall, S.H., 1993a. Pretreatment of radiata pine kraft pulp with a thermophillic enzyme Part I. Effect on conventional bleaching. Appita 46 (4), 269–273.

Allison, R.W., Clark, T.A., Wrathall, S.H., 1993b. Pretreatment of radiata pine kraft pulp with a thermophillic enzyme Part II. Effect on oxygen, ozone and chlorine dioxide bleaching. Appita 46 (5), 349–353.

Antonov, V., Maixner, V., Vicenec, R., Fishcer, H., 2005. How do enzymes contribute to bast fibres industry? In: Proceedings of the 11th Conference for Renewable Resources and Plant Biotechnology, Institute of Natural Fibres, Poznan, Poland.

Atkinson, D., Moody, D., Gronberg, V., 1993. Enzymes make pulp bleaching faster. Invest. Tech. Pap. 35 (136), 199–209.

Awakaumova, A.V., Nikolaeva, T.V., Vendilo, A.G., Kovaleva, N.E., Sinitzyn, A.P., 1999. ECF bleaching of hardwood kraft pulp: new aspects. In: 13th International papermaking conference – progress-99, Cracow, Poland, 22–24 Sept 1999, pp. IV-5-1–IV-5-13.

Babalola, T.O.O., Apata, D.F., Atteh, J.O., 2006. Effect of β-xylanase supplementation of boiled castor seed meal-based diets on the performance, nutrient absorbability and some blood constituents of pullet chicks. Trop. Sci. 46 (4), 216–223.

Baillet, E., Downey, G., Tuohy, M., 2003. Improvement of texture and volume in white bread rolls by incorporation of microbial hemicellulase preparations. In: Courtin C.M., Veraverbeke W.S.,

Delcour J.A., editors. Recent advances in enzymes in grain processing. Proceedings of the 3rd European Symposium on Enzymes in Grain Processing (ESEGP-3). Katholieke Universiteit Leuven (Leuven, Belgium): pp. 255–259.

Bajpai, P., 1997. Enzymes in pulp and paper processing. Miller Freeman, San Francisco.

Bajpai, P., 1999. Application of enzymes in pulp and paper industry. Biotechnol. Prog. 15 (2), 147–157.

Bajpai, P., 2004. Biological bleaching of chemical pulps. Crit. Rev. Biotechnol. 24 (11), 1–58.

Bajpai, P., 2006a. Advances in Recycling and Deinking. PIRA International, UK, Chapter 6, p. 75–88.

Bajpai, P., 2006b. Potential of biotechnology for energy conservation in pulp and paper. Energy management for pulp and papermakers. Budapest, Hungary, 16–18 Oct. 2006, Paper 11, 29pp.

Bajpai, P., 2009. Xylanases. In: Schaechter, M., Lederberg, J. (Eds.), Encyclopedia of Microbiology, third edition. Academic Press, San Diego, pp. 600–612.

Bajpai, P., 2012. Biotechnology in Pulp and Paper Processing. Springer-Verlag, New York.

Bajpai, P., 2013a. Pulp and paper bioprocessing. Encyclopedia of Industrial Biotechnology. John Wiley & Sons, Inc., 10.1002/9780470054581.eib297.pub2.

Bajpai, P., 2013b. Biorefinery in the Pulp and Paper Industry. Elsevier/Academic Press, Oxford.

Bajpai, P., Bajpai, P.K., 1996. Application of xylanases in prebleaching of bamboo kraft pulp. Tech. Assoc. Pulp Paper Ind. 4 (79), 225–230.

Bajpai, P., Bajpai, P.K., 1997. Realities and trends in enzymatic prebleaching of kraft pulp. Adv. Biochem. Eng. Biotechnol. 57, 1–31.

Bajpai, P., Bajpai, P.K., 1998. Deinking with enzymes: a review. Tappi J. 81 (12), 111.

Bajpai, P., Bajpai, P.K., 2001a. Status of biotechnology in pulp and paper industry. Paper International. 5(4): 29.

Bajpai, P., Bajpai, P.K., 2001b. Development of a process for the production of dissolving kraft pulp using xylanase enzyme. Appita 54 (4), 381–384.

Bajpai, P., Bhardwaj, N.K., Maheshwari, S., Bajpai, P.K., 1993. Use of xylanase in bleaching of eucalypt kraft pulp. Appita 46 (4), 274–276.

Bajpai, P., Bhardwaj, N.K., Bajpai, P.K., Jauhari, M.B., 1994. The impact of xylanases in bleaching of eucalyptus kraft pulp. J. Biotechnol. 36 (1), 1–6.

Bajpai, P., Bajpai, P.K., Kondo, R., 1999. Biotechnology for Environmental Protection in Pulp and Paper Industry. Springer, Germany, pp. 13–28 (Chapter 2).

Bajpai, P.K., Bajpai, P., Mishra, O.P., Mishra, S.P., Kumar, S., Vardhan, R., 2004. Enzyme's role in improving papermaking. 9th international conference on biotechnology in the pulp & paper industry, Durban, Sappi Forest Products, South Africa, 10–14 October 2004, p 1.10.

Bajpai, P., Bajpai, P.K., Mishra, S.P., Mishra, O.P., Kumar, S., 2005a. Enzymatic refining of pulp—case studies. In: Proceedings paperex 2005; 7th International Conference on Pulp Paper and Conversion Industry; Dec 3–5, 2005; New Delhi. pp 143–159.

Bajpai, P., Bajpai, P.K., Varadhan, R., 2005b. Production of dissolving grade pulp with hemicellulase enzyme. In: Proceedings international pulp bleaching conference; Stockholm, Sweden. pp. 303–305.

Bajpai, P., Mishra, S.P., Mishra, O.P., Kumar, S., Bajpai, P.K., 2006. Use of enzymes for reduction in refining energy—Laboratory and process scale studies. Tappi J. 5 (11), 25–32.

Basinskiene, L., Garmuviene, S., Juodeikiene, G., Haltrich, D., 2006. Fungal xylanase and its use for the bread-making process with wheat flour. World Grains Summit. Food and Beverages. San Francisco, California.

Battan, B., Dhiman, S.S., Ahlawat, S., Mahajan, R., Sharma, J., 2012. Application of thermostable xylanase of *Bacillus pumilus* in textile processing. Indian J Microbiol. 52 (2), 222–229.

Beg, Q.K., Kapoor, M., Mahajan, L., Hoondal, G.S., 2001. Microbial xylanases and their industrial applications: a review. Appl. Microbial. Biotechnol. 56, 326–338.

Biely, P., 1985. Microbial xylanolytic systems. Trends Biotechnol. 3, 286–290.

Bim, M.A., Franco, T.T., 2000. Extraction in aqueous two-phase systems of alkaline xylanase produced by *Bacillus pumilus* and its application in kraft pulp bleaching. J. Chromaogr. 743 (1), 346–349.

Bobu, E., Moraru, T., Popa, V.I., 2003. Papermaking potential improvement of secondary fibers by enzyme treatment. Cell. Chem. Technol. 37 (3–4), 305–313.

Brühlmann, F., Leupin, M., Erismann, K.H., Fiechter, A., 2000. Enzymatic degumming of ramie bast fibers. J. Biotechnol. 76, 43–50.

Butt, M.S., Tahir-Nadeem, M., Ahmad, Z., 2008. Xylanases and their applications in baking industry. Food Technol. Biotechnol. 46 (1), 22–31.

Café, M.B., Borges, A., Fritts, A., Waldroup, W., 2006. Avizyme improves performance of broilers fed corn-soybean meal-based diets. Poultry Science Department. University of Arkansas, Fayetteville AR, 72701.

Camacho, N.A., Aguilar, O.G., 2003. Production, purification and characterization of a low molecular mass xylanase from *Aspergillus* sp. and its application in bakery. Appl. Biochem. Biotechnol. 104, 159–172.

Chandrakant, P., Bisaria, V.S., 2008. Simultaneous bioconversion of cellulose and hemicellulose to ethanol. Crit. Rev. Biotechnol. 18, 295–331.

Christov, L.P., Prior, B.A., 1993. Xylan removal from dissolving pulp using enzymes of Aureobasidium pullulans. Biotechnol. Lett. 15, 1269–1274.

Christov, L.P., Prior, B.A., 1994. Enzymatic prebleaching of sulphite pulps. Appl. Microbiol. Biotechnol. 42, 492–498.

Christov, L.P., Prior, B.A., 1996. Repeated treatments with Aureobasidium pullulans hemicellulases and alkali enhance biobleaching of sulphite pulps. Enz. Microbial. Technol. 18 (4), 244–250.

Christov, L.P., Akhtar, M., Prior, B.A., 1995. Biobleaching in dissolving pulp production. In: Proceedings of the sixth international conference on biotechnology in the pulp and paper industry: advances in applied and fundamental research. Vienna, Austria: pp 625–628.

Csiszar, E., Urbanskzi, K., Szakaes, G., 2001. Biotreatment of desized cotton fabric by commercial cellulase and xylanase enzymes. J. Mol. Catal. B 11, 1065–1072.

Debyser, W., Derdelinckx, G., Delcour, J.A., 1997. Arabinoxylan solubilization and inhibition of the barley malt system by wheat during mashing with wheat whole meal adjunct, evidence for a new class of enzyme inhibitors in wheat. J. Am. Soc. Brew. Chem. 55, 153–157.

Dervilly, G., Leclercq, C., Zimmerman, D., Roue, C., Thibault, J.F., Sauliner, L., 2002. Isolation and characterization of high molecular mass water-soluble arabinoxylans from barley malt. Carbohydr. Polym. 47, 143–149.

deVries, R.P., Visser, J., 2001. Aspergillus enzymes involved in degradation of plant cell wall polysaccharides. Microbiol. Mol. Biol. Rev. 65, 497–522.

Dhiman, S.S., Sharma, J., Battan, B., 2008a. Industrial applications and future prospects of microbial xylanases: a review. Bioresources 3, 1377–1402.

Dhiman, S.S., Sharma, J., Battan, B., 2008b. Pretreatment processing of fabrics by alkalothermophilic xylanase from Bacillus stearothermophilus SDX. Enzyme Microb. Technol. 43, 262–269.

Ebbelaar, M., van der Valk, H., van Dam, J., de Jong, E., 2001. Highly efficient enzymes for the production of natural fibres, 8th International conference on biotechnology in the pulp and paper industry, Helsinki, Finland, 4–8 June 2001, p. 240.

Evans, C.T., Mann, S.P., Charley, R.C., Parfitt, D., 1995a. Formulation for treating silage containing ß-1,4-xylanase and ß-1,3-xylosidase but essentially free of ß-1,4-glucanase and ß-1,4-cellobiohydrolase, and one or more lactic acid-producing bacteria, US Patent 5432074.

Evans, C.T., Mann, S.P., Charley, R.C., Parfitt, D., 1995b. Formulation for treating silage, European Patent 0563133.

Fooks, L.J., Gibson, G.R., 2002. In vitro investigations of the effect of probiotics and prebiotics on selected human intestinal pathogens. FEMS Microbial. Ecol. 39, 67–75.

Foulk, J.A., Akin, D.E., Dodd, B.R., 2008. Pectinolytic enzymes and retting. Bioresources 3 (1), 155–169.

Fuller, J.J., Ross, R.J., Dramm, J.R., 1995. Nondestructive evaluation of honeycomb and surface checks in red oak lumber. Forest Prod. J. 45, 42–44.

Gamerith, G., Strutzenberger, H., 1992. In: Visser, J. (Ed.), Xylans and Xylanases. Elsevier, Amsterdam, pp. 339–348.

Ganga, M.A., Piñaga, F., Vallés, S., Ramón, D., Querol, A., 1999. Aroma improving in microvinification processes by the use of a recombinant wine yeast strain expressing the *Aspergillus nidulans* xlnA gene. Int. J. Food Microbiol. 47, 171–178.

Garg, N., Mahatman, K.K., Kumar, A., 2010. Xylanase: Applications and Biotechnological Aspects. Lambert Academic Publishing AG & Co. KG, Germany.

Gill, R., 2008. Advances in use of fibre modification enzymes in paper making. Conference Aticelca XXXIX Congresso Annuale, Fabriano, Italy, 29–30 May 2008.

Gilbert, H.J., Hazlewood, G.P., 1993. Bacterial cellulases and xylanases. J Gen Microbiol 139, 187–194.

Gillespie, A.M., Keane, D., Griffm, T.O., Thohy, M.G., Donaghy, J., Haylock, R.W., et al., 1990. In: Kirk, T.K., Chang, H.M. (Eds.), Biotechnology in Pulp and Paper Manufacture. Butterworth-Heinmann, Boston, pp. 211–219.

Grant, R., 1992. Enzymes reveal plenty more potential. Pulp Paper Int. 34 (9), 75–76.

Grant, R., 1993. R&D optimizes enzyme applications. Pulp Paper Int. 35 (9), 56–57.

Grant, R., 1994. Enzymes future looks bright, as range improve and expands. Pulp Paper Int. 36 (8), 20–21.

Gray, J.A., BeMiller, J.N., 2003. Bread staling: molecular basis and control. Compr. Rev. Food Sci. Food Saf. 2, 1–21.

Guy, R.C.E., Sarabjit, S.S., 2003. Comparison of effects of xylanases with fungal amylases in five flour types. Recent advances in enzymes in grain processing, Proceedings of the 3rd European Symposium on Enzymes in Grain Processing (ESEGP-3), C.M. Courtin, W.S. Veraverbeke, J.A. Delcour (Eds.), Katholieke Universiteit Leuven, Leuven, Belgium, pp. 235–239.

Hakala, T., Pursula, T., 2007. Biotechnology applications in the pulp and paper industry. In: Hermans, R., Kulvik, M., Nikinmaa, H. (Eds.), Biotechnology as a Competitive Edge for the Finnish Forest Cluster. ETLA Sarja B 227 Series. Taloustieto Oy, Helsinki, pp. 57–63. , Chapter 6.

Harbak, L., Thygesen, H.V., 2002. Safety evaluation of a xylanase expressed in *Bacillus subtilis*. Food Chem. Toxicol. 40, 1–8.

Harris, A.D., Ramalingam, C., 2010. Xylanases and its application in food industry: a review. J. Exp. Sci. 1 (7), 1–11.

Harris, G.W., Jenkins, J.A., Connerton, I., Cummings, N., Loleggio, L., Scott, M., et al., 1998. Enzymatic scouring to improve cotton fabric wettability. Text. Res. J. 68, 233–241.

Henriksson, G., Christiernin, M., Agnemo, R., 2005. Monocomponent endoglucanase treatment increases the reactivity of softwood sulphite dissolving pulp. J. Ind. Microbiol. Biotechnol. 32 (5), 211–214.

Hoondal, G.S., Tiwari, R.P., Tewari, R., Dahiya, N., Beg, Q.K., 2002. Microbial alkaline pectinases and their industrial applications: a review. Appl. Microbiol. Biotechnol. 59 (4–5), 409–418.

Izumi, Y., Kojo, A., 2003. Long-chain xylooligosaccharide compositions with intestinal function improving and hypolipemic activities, and their manufacture. Japan Patent JP 2,003,048,901.

Jiang, Z.Q., Yang, S.Q., Tan, S.S., Li, L.T., Li, X.T., 2005. Characterization of a xylanase from the newly isolated thermophilic thermomyces lanuginosus CAU44 and its application in bread making. Lett. Appl. Microbiol. 41, 69–76.

Juturu, V., Wu, J.C., 2012. Microbial xylanases: engineering, production and industrial applications. Biotechnol. Adv., 2012. Available from: http://dx.doi.org/10.1016/j.bbr.2011.03.031.

Kantelinen, A., Hortling, B., Sundquist, J., Linko, M., Viikari, L., 1993. Proposed mechanism of the enzymatic bleaching of kraft pulp with xylanases. Holzforschung 47, 318–324.

Karmakar, S.R., 1999. Textile science and technology series. 1. Amsterdam: Elsevier Science B.V. Chemical technology in the pretreatment processes of textiles. p. 12.

Kazumitsu, S., Boseki, I., Norio, S., Yoshimasa, O., 1987. Production of food and drink. Japan Patent JP 62278961;1987.

Kazumitsu, S., Boseki, I., Norio, S., Yoshimasa, O., 1997. Production of food and drink. Japan Patent JP 9248153;1997.

Kenealy, W.R., Jeffries, T.W., 2003. Enzyme processes for pulp and paper: a review of recent developments. Wood deterioration and preservation: advances in our changing world. In: Keskin, S.O., Sumnu, G., Sahin, S. (Eds.), Usage of Enzymes in a Novel Baking Process. Nahrung/Food, 48. pp. 156–160.

Ko, C.-H., Tsai, C.-H., Tu, J., Yang, B.-Y., Hsieh, D.-L., Jane, W.-N., et al., 2011. Identification of Paenibacillus sp. 2 S-6 and application of its xylanase on biobleaching. Int. Biodeterior Biodegr. 65 (2), 334–339.

Kuhad, R.C., Singh, A., 1993. Lignocellulosic biotechnology: current and future prospects. Crit. Rev. Biotechnol. 13, 151–172.

Laurikainen, T., Haerkoenen, H., Autio, K., Poutanen, K., 1998. Effects of enzymes in fibre enriched baking. J. Sci. Food Agric. 76, 239–249.

Loosvelt, I., 2009. Current applications of fibre modification enzymes in the paper industry and future possibilities. Fibre Engineering, Gothenburg, Sweden, p 39.

Losonczi, A., Csiszar, E., Szakacs, G., Bezur, L., 2005. Role of the EDTA chelating agent in bioscouring of cotton. Text Res. J. 75 (5), 411–417.

Maat, J., Roza, M., Verbakel, J., Stam, H., Santos da Silva, M.J., Bosse, M., et al., 1992. Xylanases and their applications in bakery. In: Visser, J., Beldman, G., Someren Kusters-van, M. A., Voragem, A.G.J. (Eds.), Xylans and Xylanases. Elsevier, Amsterdam, pp. 349–360.

Manji, A.H., 2006. Extended usage of xylanase enzyme to enhance the bleaching of softwood kraft pulp. TAPPI J. 5 (1), 23–26.

Martinez-Anaya, M.A., Jimenez, T., 1997. Functionality of enzymes that hydrolyse starch and nonstarch polysaccharide in breadmaking. Z Lebensm Unters Forsch 205, 209–214.

Mathewson, P.R., 2000. Enzymatic activity during bread baking. Cereal Food World 45, 98–101.

Mathlouthi, N., Saulnier, L., Quemener, B., Larbier, M., 2002. Xylanase, beta-glucanase, and other side enzymatic activities have greater effects on the viscosity of several feedstuffs than xylanase and beta-glucanase used alone or in combination. J. Agric. Food Chem. 50, 5121–5127.

Minor, J.L., 1986. Chemical linkage of polysaccharides to residual lignin in loblolly pine kraft pulps. J. Wood Chem. Technol. 6 (2), 185–201.

Modler, H.W., 1994. Bifidogenic factors: sources, metabolism and applications. Int. Dairy J. 4, 383–407.

Monfort, A., Blasco, A., Prieto, J.A., Sanz, P., 1997. Construction of baker's yeast strains that secrete different xylanolytic enzymes and their use in bread making. J. Cereal Sci. 26, 195–199.

Novozymes, 2013. Panzea. Available at: http://www.novozymes.com/en/solutions/food-and-beverages/baking/bread-and-rolls/doughimprovement/Panzea/Pages/default.aspx

Ojanpera, K., 2004. Biotechnology breaks through to forest industry. Tek. Talous2, no. 17.

Olsson, L., Hahn-Hägerdal, B., 1996. Fermentation of lignocellulosic hydrolysates for ethanol production. Enzyme Microb. Technol. 18, 312–331.

Ozmutlu, O., Sumnu, G., Sahin, S., 2001. Effects of different formulations on the quality of microwave baked breads. Eur. Food Res. Technol. 213, 38–42.

Paice, M., Zhang, X., 2005. Enzymes find their niche. Pulp Paper Can. 106 (6), 17–20.

Paice, M.G., Gurnagul, N., Page, D.H., Jurasek, L., 1992. Mechanism of hemicellulose directed prebleaching of kraft pulp. Enzyme Microb. Technol. 14, 272–276.

Parajó, J.C., Domíngues, H., Domíngues, J.M., 1998. Biotechnological production of xylitol. Part 1, interest of xylitol and fundamentals of its biosynthesis. Bioresour. Technol. 65, 191–201.

Pérez, J., Muñoz-Dorado, J., de la Rubia, T., Martínez, J., 2002. Biodegradation and biological treatments of cellulose, hemicellulose and lignin: an overview. Int. Microbiol. Off. J. Spanish Soc. Microbiol. 5, 53–63.

Polizeli, M.L., Rizzatti, A.C., Monti, R., Terenzi, H.F., Jorge, J.A., Amorim, D.S., 2005. Xylanases from fungi: properties and industrial applications. Appl. Microbiol. Biotechnol. 67, 577–591.

Prade, R.A., 1995. Xylanases, from biology to biotechnology. Biotechnol. Genet. Eng. Rev. 13, 101–131.

Ratto, M., Kantelinen, A., Bailey, M., Viikari, L., 1993. Potential of enzymes for wood debarking. Tappi J. 76 (2), 125–128.

Redgwell, R.J., de Michieli, J.H., Fischer, M., Reymond, S., Nicholas, P., Sievert, D., 2001. Xylanase induced changes to water- and alkali- extractable arabinoxylans in wheat flour: Their role in lowering batter viscosity. J. Cereal Sci. 33, 83–96.

Reeve, D.W., 1996a. Introduction to the principles and practice of pulp bleaching. In: Dence, C. W., Reeve, D.W. (Eds.), Pulp Bleaching: Principles and Practice. Tappi Press, Atlanta, p. 1., Section 1, Chapter 1.

Reeve, D.W., 1996b. Pulp bleaching: principles and practice. In: Dence, C.W., Reeve, D.W. (Eds.), Chlorine Dioxide in Bleaching Stages. Tappi Press, Atlanta, Section 4, Chapter 8, p. 379.

Reid, I.D., Bourbolnnais, R., Paice, M.G., 2010. Biopulping and biobleaching. In: Heitner, C., Dimmel, D.R., Schmidt, J.A. (Eds.), Lignin and Lignans: Advances in Chemistry. CRC Press, Boca Raton (FL), pp. 521–554.

Rouau, X., El-Hayek, M.L., Moreau, D., 1994. Effect of an enzyme preparation containing pentosanases on the bread-making quality of flours in relation to changes in pentosan properties. J. Cereal Sci. 19, 259–272.

Rouette, H.K., 2001. Encyclopedia of Textile Finishing. Springer, Berlin, pp. 1–3.

Rycroft, C.E., Jones, M.R., Gibson, G.R., Rastall, R.A., 2001. A comparative in vitro evaluation of the fermentation properties of prebiotic oligosaccharides. J. Appl. Microbiol. 91, 878–887.

Saha, B., Bothast, R.J., 1997. Microbial production of xylitol. In: Saha, B.C., Woodward, J. (Eds.), Fuels and Chemicals from Biomass. American Chemical Society, Washington, DC, pp. 307–319.

Saha, B.C., Bothast, R.J., 1999. Enzymology of xylan degradation. In: Imam, S.H., Greene, R. V., Zaidi, B.R. (Eds.), Biopolymers: utilizing natures advanced materials. American Chemical Society, Washington, DC, pp. 167–194.

Saleem, M., Tabassum, M.R., Yasmin, R., Imran, M., 2009. Potential of xylanase from thermophilic Bacillus sp. XTR-10 in biobleaching of wood kraft pulp. Int. Biodeterior. Biodegr. 33 (8), 1119–1124.

Screenath, H.K., Jeffries, T.W., 2000. Production of ethanol from wood hydrolysate by yeasts. Bioresour. Technol. 72, 253–260.

Senior, D.J., Hamilton, J., 1991. Use of xylanase to decrease the formation of AOX in kraft pulp bleaching. In: Proceedings of the environment conference of the technical section, Canadian Pulp and Paper Association, Quebec, Canada, 8–10 Oct 1991, pp. 63–67.

Senior, D.J., Hamilton, J., 1992a. Bleaching with xylanases brings biotechnology to reality. Pulp Pap. 66 (9), 111–114.

Senior, D.J., Hamilton, J., 1992b. Reduction in chlorine use during bleaching of kraft pulp following xylanase treatment. TAPPI J. 75 (11), 125–130.

Senior, D.J., Hamilton, J., 1992c. Use of xylanase to decrease the formation of AOX in kraft pulp bleaching. J. Pulp Pap. Sci. 18 (5), J165–J168.

Senior, D.J., Hamilton, J., 1993. Xylanase treatment for the bleaching of softwood kraft pulps: the effect of chlorine dioxide substitution. TAPPI J. 76 (8), 200–206.

Senior, D.J., Hamilton, J., Bernier Jr, R.L., 1992. Use of Streptomyces lividans xylanase for biobleaching of kraft pulps. In: Visser, J., Beldmann, G., Kusters-van Someren, M.A., Voragen, A.G.J. (Eds.), Xylans and Xylanases. Progress in Biotechnology, vol. 7. Elsevier, Amsterdam, p. 555.

Senior, D.J., Hamilton, J., Taiplus, P., Torvinen, J., 1999. Enzyme use can lower bleaching costs, aid ECF conversions. Pulp Pap. 73 (7), 59–62.

Senior, D.J., Bernhardt, S.A., Hamilton, J., Lundell, R., 2000. Mill implementation of enzymes in pulp manufacture. In: Biological science symposium, San Francisco, CA, 19–23 Oct 2000, pp 163.

Senn, T., Pieper, H.J., 2001. The biotechnology of ethanol. In: Roeher, M. (Ed.), Classical and Future Applications, 48. Wiley-VCH, Germany, pp. 8–174.

Sharma, H., 1987. Screening of polysaccharide-degrading enzymes for retting flax stem. Int. Biodeterior. 23 (3), 181–186.

Sharma, M., Kumar, A., 2013. Xylanases: an overview. Brit. Biotechnol. J. 3 (1), 1–28.

Shatalov, A., Pereira, H., 2009. Impact of hexenuronic acid on xylanase aided biobleaching of chemical pulps. J. Bioresour. Technol. 100, 3069–3075.

Skerker, P.S., Labbauf, M.M., Farrell, R.L., Beerwan, N., McCarthy, P., 1992. Practical bleaching using xylanases: laboratory and mill experience with Cartazyme HS-10 in reduced and chlorine free bleach sequences. In: TAPPI pulping conference, Boston, 1–5 Nov 1992. TAPPI Press, Atlanta, GA, p. 27.

Skjold-Jorgensen, S., Munk, N., Pederson, L.S., 1992. Recent progress within the application of xylanases for boosting the bleachability of kraft pulp. In: Kuwahara, M., Shimada, M. (Eds.), Biotechnology in the Pulp and Paper Industry. Uni Publishers, Tokyo, pp. 93–99.

Smook, G.A., 1992. Wood and Chip Handling. Handbook for Pulp & Paper Technologists, second ed. Angus Wilde Publications, Vancouver, p 20.

Sorensen, J.F., Sibbesen, O. and Poulsen, C.H., 2001. Degree of inhibition by the endogenous wheat xylanase inhibitor controls the functionality of microbial xylanases 'in Dough'. AACC Annual Meeting, Enzymes and Baking – 213AB, Charlotte, NC, USA.

Spagna, G., Ramagnoli, D., Angela, M., Biochi, G., Pifferi, P.G., 1998. A simple method for purifying glycosidase α-Larabinofuranosidase and β-D-glucopyranosidase from *A. niger* to increase the aroma of wine. I. Enzyme Microb. Technol. 22, 298−304.

Suurnakki, A., Tenkanen, M., Buchert, J., Viikari, L., 1997. Hemicellulases in the bleaching of chemical pulps. In: Scheper, T. (Ed.), Advances in Biochemical Engineering/Biotechnology, vol. 57. Springer, Berlin, pp. 261−287.

Taeko, I., Koichi, N., Yasushi, N., Akiraand, K., Yoshinobu, K., 1998. Food and drink effective in anti-obesity. Japan Patent JP 10290681.

Tan, S.S., Li, D.Y., Jiang, Z.Q., Zhu, Y.P., Shi, B., Li, L.T., 2008. Production of xylobiose from the autohydrolysis explosion liquor of corncob using Thermotogamaritima xylanase B (XynB) immobilized on nickel-chelated Eupergit C. Bioresour. Technol. 99, 200−204.

Thibault, L., Tolan, J., White, T., Yee, E., April, R., Sung, W., 1999. Use of an engineered xylanase enzyme to improve ECF bleaching at Weyenhaeuser Prince Albert. In: 85th Annual meeting, Montreal, QC, 26−29 Jan 1999, pp B263.

Tolan, J.S., 1992. Mill implementation of enzyme treatment to enhance bleaching. In: Proceedings of the 78th CPPA annual meeting, Montreal, QC, 28−29 Jan 1992, pp. A163−A168.

Tolan, J.S., 2001. How a mill can get more benefit out of its xylanase treatment. In: 8th International conference on biotechnology in the pulp and paper industry, Helsinki, Finland, 4−8 June 2001, pp. 81.

Tolan, J.S., Canovas, R.V., 1992. The use of enzymes to decrease the chlorine requirements in pulp bleaching. Pulp Pap. Can. 93 (5), 39−42.

Tolan, J.S., Guenette, M., 1997. Using enzymes in pulp bleaching: mill applications. In: Scheper, T. (Ed.), Advances in Biochemical Engineering/Biotechnology, vol. 57. Springer, Berlin, pp. 288−310.

Tolan, J.S., Guenette, M., Thebault, L., Winstanley, C., 1994. The use of a novel enzyme treatment to improve the efficiency of shive removal by bleaching. Pulp Pap. Can. 95 (12), T488.

Tolan, J.S., Olson, D., Dines, R.E., 1996. Survey of mill usage of xylanase. In: Jeffries T.W., Viikari L., editors. Enzymes for Pulp and Paper Processing. ACS Symposium Series 655, American Chemical Society, Washington, DC, pp 25−35.

Toshio, I., Noriyoshi, I., Toshiaki, K., Toshiyuki, N., Kunimasa, K., 1990. Production of Xylobiose. Japan Patent JP 2119790.

Twomey, L.N., Pluske, J.R., Rowe, J.B., Choct, M., Brown, W., McConnell, M.F., et al., 2003. The effects of increasing levels of soluble nonstarch polysaccharides and inclusion of feed enzymes in dog diets on faecal quality and digestibility. Anim. Feed Sci. Technol. 108 (1−4), 71−82.

US Department of Energy. 2012. Xtreme Xylanase Discovery Aims to Revolutionize Biorefineries. Idaho National Laboratory. Available at: http://www.inl.gov/research/xtreme-xylanase/#sthash.yYG1K8Fb.dpuf

Valchev, V., Valchev, V., Christova, E., 1998. Introduction of an enzyme stage in bleaching of hardwood kraft pulp. Cell. Chem. Technol. 32 (5−6), 457−462.

Valchev, I., Valchev, I., Ganev, I., 2000. Improved elemental chlorine free bleaching of hardwood kraft pulp. Cell. Chem. Technol. 33 (1−2), 61−66.

Valls, C., Vidal, T., Roncero, M.B., 2010. The role of xylanases and laccases on hexenuronic acid and lignin removal. Process Biochem. 45, 425−430.

Van Gorcom, R.F.M., Hessing, J.G.M., Maat, J., Roza, M., Verbakel, J.M.A., 2003. Xylanase production, US Patent 6586209.

Van Sumere, C.F., 1992. Retting of flax with special reference to enzyme-retting. In: Sharma, H.S.S., Van Sumere, C.F. (Eds.), The Biology and Processing of Flax. M Publications, Belfast, pp. 157−198.

Vazquez, M.J., Alonso, J.L., Dominguez, H., Parajo, J.C., 2000. Xylooligosaccharides: manufacture and applications. Trends Food Sci. Technol. 11, 387–393.

Verschraege, L., 1989. Cotton Fibre Impurities. Neps, Motes and Seed Coat Fragments. ICAC Review Articles on Cotton Production Research No 1. CAB International, Wallingford.

Viikari, L., Ranua, M., Kantelinen, A., Sundquist, J., Linko, M.1986. Bleaching with enzymes. In: Proceedings of the 3rd international conference on biotechnology in the pulp and paper industry, Stockholm, Sweden, pp 67–69.

Viikari, L., Rato, M., Kantelinen, A., 1989. Finnish Patent Appl 896291.

Viikari, L., Kantelinen, A., Poutanen, K., Ranua, M., 1990. Characterization of pulps treated with hemicellulolytic enzymes prior to bleaching. In: Kirk, T.K., Chang, H.M. (Eds.), Biotechnology in Pulp and Paper Manufacture. Butterworth-Heinemann, Boston, p. 145.

Viikari, L., Kantelinen, A., Rättö, M., Sundquist, J., 1991. Enzymes in pulp and paper processing. In: Leatham, G.F., Himmel, M.E. (Eds.), Enzymes in Biomass Conversion, ACS Symp Ser 460. American Chemical Society, Washington, pp. 426–436.

Viikari, L., Tenkanen, M., Buchert, J., Ratto, M., Bailey, M., Siika-aho, M., et al., 1993. Hemicellulases for industrial applications. In: Saddler, J.N. (Ed.), Bioconversion of Forest and Agricultural Plant Residues. CAB International, Wallingford (UK), pp. 131–182.

Viikari, L., Kantelinen, A., Sundquist, J., Linko, M., 1994. Xylanases in bleaching: from an idea to industry. FEMS Microbial. Rev. 13, 335–350.

Viikari, L., Poutanen, K., Tenkanen, M., Tolan, J.S., 2002. Hemicellulases. In: Flickinger, M.C., Drew, S.W. (Eds.), Encyclopedia of Bioprocess Technology: Fermentation, Biocatalysis, and Bioseparation. Wiley, Chichester, West Sussex (Update. Electronic release).

Viikari, L., Suurna kki, A., Gronqvist, S., Raaska, L., Ragauskas, A., 2009. Forest products: biotechnology in pulp and paper processing. In: Schaechter, M. (Ed.), Encyclopedia of Microbiology, third ed. Academic Press, New York, pp. 80–94.

Viikari, L., Vehmaanperä, J., Koivula, A., 2012. Lignocellulosic ethanol: from science to industry. Biomass Bioenergy, 1–12.

Wang, M., van Vliet, T., Hamer, R.J., 2004. Evidence that pentosans/ xylanase affects the re-agglomeration of the gluten network. J. Cereal Sci. 39, 341–349.

Wang, S. and Kim, M., 2005. Study on old newsprint deinking with cellulases and xylanase, 59th Appita Annual Conference and Exhibition, ISWFPC, Auckland.

Werthemann, D., 1993. Prebleaching of Pinus radiata pulp using enzymes – technology to reduce AOX. Jpn. J. Pap. Technol. 10, 15–17.

Whitehead, T.R., Cotta, M.A., 2001. Identification of a broad-specificity xylosidase/arabinosidase important for xylooligosaccharide fermentation by the ruminal anaerobe Selenomonas ruminantium GA 192. Curr. Microbiol. 43, 293–298.

Winkelhausen, E., Kuzmanova, S., 1998. Microbial conversion of D-xylose to xylitol. J. Ferment Bioeng. 86, 1–14.

Wong, K.K.Y., Tan, L.U.L., Saddler, J.N., 1988. Multiplicity of β-1,4-xylanase in microorganisms: functions and applications. Microbiol Rev. 52, 305–317.

Wong, K.K.Y., Saddler, J.N., 1993. Applications of hemicellulases in the food, feed, and pulp and paper industries. In: Coughlan, M.P., Hazlewood, G.P. (Eds.), Hemicelluloses and Hemicellulases. Portland Press, London, pp. 127–143.

Wong, K.K.Y., Allison, R.W., Spehr, S., 2001. Effect of alkali and oxygen extractions of kraft pulps on xylanase aided bleaching. J. Pharm. Pharmaceutical Sci., 7.

Yang, J.L., Eriksson, K.-E.L., 1992. Use of hemicellulolytic enzymes as one stage in bleaching of kraft pulps. Holzforschung 46 (6), 481–488.

Yang, R., Xu, S., Wang, Z., Yang, W., 2005. Aqueous extraction of corncob xylan and production of xylooligosaccharides. LWT—Food Sci. Technol. 38, 677–682.

Yllner, S., Ostberg, K., Stockmann, L., 1957. A study of the removal of the constituents of pine wood in the sulphate process using a continuous liquor flow method. Sven. Papperstidn. 60, 795–802.

Zhan, H., Yue, B., Hu, W., Huang, W., 2000. Kraft reed pulp TCF bleaching with enzyme treatment. Cell. Chem. Technol. 33 (1–2), 53–60.

Actinomycetes Actinomycetes are prokaryotic and heterotrophic organisms that belong to the soil microflora (<0.2 mm). They are single-celled, filamentous, and often profoundly branched organisms.

Aerobic A process that occurs in the presence of air or oxygen.

Algae Algae are simple-celled plants and (like all plants) contain chlorophyll. This traps energy from the sun and uses that energy to convert nutrients and carbon dioxide (which are dissolved in the water) into growth. When grown in a hatchery, the growing and multiplying algal cells are collectively known as a culture. The main form of algae that is of interest to aquaculture is collectively known as unicellular algae.

Anaerobic Conditions where oxygen is lacking; organisms not requiring oxygen for respiration.

Angiosperm A seed-bearing plant with the ovules borne enclosed by a megasporophyll whose margins are variously fused (i.e., the ovules are enclosed in an ovary).

Arabinan Arabinan is a common hemicellulosic component of the cell wall of higher plants. It is a branched polysaccharide with a backbone of α-(1,5)-linked arabinofuranosyl residues with α-(1,2)- and α-(1,3)-linked arabinofuranosyl side chains.

Bacteria Unicellular, prokaryotic, microscopic, generally heterotrophic organisms present in great numbers in soil and in water; largely responsible for decomposition of primary and secondary produced organic matter and for mineralization of its constituent elements: C, N, P, S, etc. The bacteria are one of the three domains of life, the other being Archaea and Eukarya (eukaryotes).

Bioethanol Ethanol produced from biomass and/or the biodegradable fraction of waste.

Biological oxygen demand A measure of the amount of oxygen consumed in biological processes that breakdown organic matter in water. The greater the BOD, the higher the degree of pollution.

Biomass All materials that are produced by photosynthesis and are potentially useful for the production of organic chemicals or as energy sources.

Biorefinery A biorefinery is a facility that integrates biomass conversion processes and equipment to produce fuels, power, and value-added chemicals from biomass. The biorefinery concept is analogous to today's petroleum refinery, which produces multiple fuels and products from petroleum.

Black liquor The liquor that exits the digester with the cooked chips at the end of the kraft cook. It is a mixture of the cooking chemicals and dissolved wood material that remains after sulphate cooking; it is recovered during pulp washing, concentrated by evaporation, and burned in the recovery boiler to regenerate the cooking chemicals and energy.

Bleaching The process of brightening the fiber by removal or decolorization of the colored substance.

Brightness The reflectance or brilliance of the paper when measured under a specially calibrated blue light. Not necessarily related to color or whiteness, brightness is expressed in %ISO.

Brownstock The suspension of unbleached pulp.

Carbohydrates Large group of polymer compounds synthesized by plants containing carbon, hydrogen, and oxygen, in which the latter two elements are usually in the 2:1 proportion of water. Cellulose, sugars, and starches are all carbohydrates.

Cellulases A family of enzymes that hydrolyze β-1,4-glucosidic bonds in native cellulose and derived substrates.

Cellulose A high molecular weight, linear polymer of repeating beta-D-glucopyranose units. It is the chief structural element and major constituent of the cell wall of trees and plants.

Cellulosomes Cellulosomes are multi-enzyme complexes associated with the cell surface that mediate cell attachment to the insoluble substrate and degrade it to soluble products that are then absorbed. Cellulosome complexes are intricate multi-enzyme machines produced by many cellulolytic microorganisms.

Cell wall The rigid outermost cell layer found in plants and certain algae, bacteria, and fungi, but characteristically absent from animal cells.

Chelating agent A multidentate ligand. It simultaneously attaches to two or more positions in the coordination sphere of a central metal ion. Chelants or chelating agents such as ethylene-diamine-tetraacetic acid (EDTA) and diethylene-triamine-penta-acetic acid (DTPA) are applied because of good sequestering properties (i.e., their ability to suppress the activity of dissolved transition metal ions without precipitation).

Chemical pulp A generic term that describes pulp produced by chemical (as opposed to mechanical) processes. These chemical processes include Kraft (sulphate) and sulphite processes.

Consistency The weight percent of air dry (or oven dry) fibrous material in a stock or stock suspension. Typical ranges are: low consistency (3−5%, LC), medium consistency (10−15%, MC), and high consistency (30−50%, HC).

Constitutive enzyme An enzyme whose formation is not dependent on the presence of a specific substrate.

Contaminant Biological (e.g., bacterial and viral pathogens) and chemical introductions capable of producing an adverse response (effect) in a biological system, seriously injuring structure or function or producing death.

Deinking The process of removing ink from printed wastepapers, but also involving general removal of other undesirable materials.

Delignification The removal of lignin—the material that binds wood fibers together—during the chemical pulping process.

Desizing Desizing is the process of removing sizing materials from the fabric, which is applied in order to increase the strength of the yarn so it can withstand the friction of the loom. Fabric that has not been desized is very stiff and causes difficulty in its treatment with different solutions in subsequent processes.

Dioxins The term used to describe the families of chemicals known as chlorinated dibenzo-p-dioxins and dibenzo-p-furans. These families consist of 75 chlorinated dibenzo-p-dioxins and 135 chlorinated dibenzo-p-furans. These molecules can have from one to eight chlorine atoms attached to a planar carbon skeleton. 2,3,7,8-tetrachloro-dibenzo-p-dioxin (TCDD) and 2,3,7,8-tetrachlorodibenzofuran (TCDF) are two of the most toxic members of this family of compounds. If dioxins are detected in releases from bleaching processes that expose unbleached pulp to elemental chlorine, the dioxins are most likely to be TCDD and TCDF.

Effluent A complex waste material (e.g., liquid industrial discharge or sewage) that may be discharged into the environment.

Elemental chlorine free (ECF) A bleaching process that uses chlorine dioxide as opposed to elemental chlorine gas. ECF papers are made exclusively from pulp that uses chlorine dioxide rather than elemental chlorine gas as a bleaching agent.

Enzyme A substance containing protein that catalyzes biological reactions.

***Eucalyptus* spp.** The genus *Eucalyptus* belongs to the family Myrtaceae. Worldwide, Myrtaceae comprises some 140 genera and 3000 species (spp.).

Fungi A kingdom of life-forms that are eukaryotic, mycelial, or yeast-like, heterotrophic, lacking in chlorophyll, sexually and/or asexually reproductive, and mostly aerobic.

Furfural Furfural is an organic compound derived from a variety of agricultural by-products.

Galactan Polysaccharide composed of repeating galactose units. They can consist of branched or unbranched chains in any linkages.

Gymnosperm A seed-bearing plant with the ovules borne on the margins or surface of a sporophyll and not enclosed by fusion of the sporophyllar tissue; a vascular plant having seeds that are not enclosed in an ovary.

Hardwood Pulpwood from broad-leaved, dicotyledonous, deciduous trees, such as aspen, beech, birch, and eucalyptus.

Hemicellulases Enzymes that hydrolyze arabinogalactans, galactans, xyloglucans, and xylans.

Hemicelluloses Short-chain polysaccharides having a DP (degree of polymerization) of 15 or less; mainly polymers of sugars other than glucose. Principal hemicelluloses are xylan in hardwoods and gluco-mannan in softwoods.

Hexenuronic acids Approximately 75% of the hemicelluloses in hardwoods consist of xylan (4-O-methyl-glucuronoarabinoxylan). Under the kraft pulping conditions, xylan generates hexenuronic acid groups (HexA) that form a considerable part of the residual oxidizable material after pulping (measured by the Kappa number). Hardwood kraft pulps, and especially eucalypt kraft pulps, may contain high amounts of hexenuronic acids (HexA), contrary to softwood kraft pulps. HexA have adverse effects in bleaching.

The most important ones are: increased consumption of bleaching agents, such as chlorine dioxide (ClO_2) and ozone (O_3), to reach target brightness; increased brightness reversion; and contribution to formation and scaling of oxalates in bleaching equipment.

Immobilized enzyme An immobilized enzyme is an enzyme that is attached to an inert, insoluble material. This can provide increased resistance to changes in conditions such as pH or temperature. It also allows enzymes to be held in place throughout the reaction, after which they are easily separated from the products and may be used again—a far more efficient process, so it is widely used in industry for enzyme-catalyzed reactions. An alternative to enzyme immobilization is whole cell immobilization.

Kappa number A measure of residual lignin content in unbleached pulp, determined after pulping and prior to bleaching. The lower the Kappa number, the less associated the lignin.

Kraft Pulp A chemical pulp produced by combining wood chips and chemicals in huge vats known as digesters. The effect of the heat and the chemicals dissolves the lignin that binds the cellulose fibers together, without breaking the wood fibers, creating a strong pulp product.

Lignin Structural constituent of wood and (to a lesser extent) other plant tissues, which encrusts the cell walls and cements the cells together. It is not fermentable.

Mannan Polysaccharide in which mannose is the main sugar.

Mechanical pulp Mass of fibers separated from wood chips by mechanical energy during refining. It contains cellulose, hemicelluloses, and lignin.

Mesophile An organism that grows best in the temperature range of 20°C to 50°C; optimal growth often occurs at about 37°C.

Messenger RNA (mRNA) It is a single-stranded RNA molecule that is complementary to one of the DNA strands of a gene. The mRNA is an RNA version of the gene that leaves the cell nucleus and moves to the cytoplasm where proteins are made. During protein synthesis, an organelle called a ribosome moves along the mRNA, reads its base sequence, and uses the genetic code to translate each three-base triplet, or codon, into its corresponding amino acid.

Middle lamella The pectin-rich intercellular material cementing together the primary walls of adjacent plant cells.

Molecular biology The study of biological phenomena at the level of molecule, including techniques with which genes can be purified, sequenced, changed, and introduced into cells. It provides an integrated experimental approach to problems in genetic, biochemistry, and prokaryotic and eukaryotic cell biology.

Monosaccharides The simplest repeating units of carbohydrates. Also known as sugars or saccharides (glucose and fructose are monosaccharides).

Mutant An organism with a changed or new gene.

Oligosaccharides Carbohydrates that hydrolyze to yield 2−10 molecules of a monosaccharide (maltose and sucrose are disaccharides).

Paper Sheet of fibers with a number of added chemicals. According to the basic weight, it can be distinguished as follows: Paper < 150 g/m^2 $<$ paper-board (or board) < 250 g/m^2 $<$ cardboard.

Pathogen An agent that causes disease, especially a living microorganism such as a bacterium or fungus.

Pectin A highly hydrophilic polysaccharide built up of monomers of galacturonic acid; an important component of cell walls.

Pectinase Pectinase, also known as polygalacturonase, is the collective term for a row of enzymes that are able to break down or to transform pectins.

Perennial plant A perennial plant is one that will last for several years. Perennials will survive winter and return with new growth during the onset of the growing season.

Pollution Generally, the presence of matter or energy whose nature, location, or quantity produces undesired environmental effects.

Polysaccharides Carbohydrates that hydrolyze to yield more than 10 molecules of a monosaccharide (cellulose and starch are glucose polymers).

Prebiotics Prebiotics are ingredients that stimulate the growth and/or function of beneficial intestinal microorganisms.

Primary treatment Physical treatment of wastewater to reduce settleable and floatable solids.

Protein engineering The design, development, and production of new protein products having properties of commercial value.

Protozoa Large group of unicellular animals that are bigger and more complex than bacteria.

Pulping The process of converting raw fiber (e.g., wood) or recycled fiber to a pulp usable in papermaking.

Reactivity Reactivity may be defined in terms of the fraction of the residual oxidizable material that a bleaching agent is capable of removing (i.e., delta kappa/delta bleaching agent).

Recombinant DNA technology Uses enzymes to cut and paste together DNA sequences of interest. The recombined DNA sequences can be placed into vehicles called vectors that ferry the DNA into a suitable host cell where it can be copied or expressed.

Refining Process of mechanically treating fibers to develop strength.

Saprotroph Any organism, especially a fungus or bacterium, that lives and feeds on dead organic matter.

Scouring Scouring is a chemical washing process carried out on cotton fabric to remove natural wax and nonfibrous impurities from the fibers along with any added soiling or dirt.

Softwood Wood obtained from evergreen cone-bearing species of trees—such as pines, spruces, etc.—that are characterized by having needles.

Starch Starch is a polysaccharide used by plants to stockpile glucose molecules. It is the major component of flour, potatoes, rice, beans, corn, and peas. Starch is a mixture of two different polysaccharides: amylose (about 20%), which is insoluble in cold water, and amylopectin (about 80%), which is soluble in cold water. Amylose is composed of unbranched chains of D-glucose units joined by $\alpha(1 \rightarrow 4)$-glycosidic linkages. Unlike amylose, which is a linear polymer, amylopectin contains $\alpha(1 \rightarrow 6)$-glycoside branches approximately every 25 glucose units.

Synergism Two or more factors acting cooperatively, so that their combined effects when acting together exceed the sum of their effects when each acts alone.

Thermomechanical pulp (TMP) A high-yield pulp produced from wood chips that have been exposed under pressure to superheated steam. The heat softens the lignin, which allows fiber separation with less damage than in purely mechanical pulping.

Thermophile An organism that grows at a higher temperature than most other organisms. In general, a wide range of bacteria, fungi, and simple plants and animals can grow at temperatures up to 50°C; thermophiles are considered to be organisms that can grow at above 50°C. They can be classified according to their optimal growth temperatures, into: simple thermophiles (50−65°C), thermophiles (65−85°C), and extreme thermophiles (>85°C). Thermophiles and extreme thermophiles are usually found growing in very hot places, such as hot springs and geysers, smoker vents on the seafloor, and domestic hot water pipes.

Totally chlorine free (TCF) Bleaching process that uses no chlorine products.

Viscosity The property of a fluid that resists the force tending to cause the fluid to flow.

Xylans A polysaccharide component of hemicellulose containing only xylose molecules.

Xylanosomes These are discrete, multifunctional, multienzyme complexes found on the surface of several microorganisms.

Xylitol Sugar alcohol derived from xylose.

Xylose A five-carbon sugar with an aldehyde functional group, common in hemicelluose.

Yeast Yeasts are a group of unicellular fungi that exist almost everywhere in nature. Commonly used to leaven bread and ferment alcoholic beverages.